Small Signal
Microwave Amplifier Design

Small Signal
Microwave Amplifier Design

Theodore Grosch

NOBLE
PUBLISHING

Noble Publishing Corporation
Atlanta, GA

Library of Congress Cataloging-in-Publication Data

Grosch, Theodore, 1957-
    Small signal microwave amplifier design / Theodore Grosch.
        p. cm.
    Includes index.
    ISBN 1-884932-06-1
        1. Microwaves amplifiers--Design and construction. 2. Electronic circuit design. I.
Title.

TK7871.2 .G76 1999
621.381'325--dc21

                                                                99-088646

NOBLE
PUBLISHING

ISBN 1-884932-06-1

To Mary, Douglas and Caroline

Contents

Acknowledgements

I thank Rose Pruyne who did an excellent job editing the manuscript. Her work was invaluable in the production of the text. I also thank Dr. Lynn Carpenter who reviewed this book and made several import corrections and comments.

Preface

This book focuses on analytical methods of high-frequency amplifier design by determining the characteristics of input and output networks and their subsequent synthesis. These techniques are combined into a methodology for designing narrowband, small signal amplifiers at RF and microwave frequencies. Applying these techniques facilitates the follow-on stages of modeling and testing the amplifier. Topics covered in this book include:

- Introduction to Network Parameters
- Transmission Lines
- Introduction to the Smith Chart
- S-Parameters
- Introduction to Microstrip Lines
- Circuit Analysis Using S-Parameters
- Synthesis-Matching Circuits Using Lumped Elements and Transmission Lines
- Circuit Stability
- Narrowband Amplifier Design
- Amplifier Gain and Constant Gain Circles
- Introduction to Broadband Amplifier Design
- Noise Figure Basics and Low Noise Amplifier Design
- Introduction to Noise Measurements.

Each section includes an example showing how to apply the technique discussed. Additional problems are given at the end of each chapter. For those with access to MatLab by The Math Works or Genesys by Eagleware, program listings are included.

Before amplifier design can be discussed, basic tools and techniques for high-frequency circuit design have to be given. The most important tools for microwave circuit design are S-parameters and the Smith chart. We begin by reviewing the basic concepts of networks and their definition, i.e., the common network parameters of impedance, admittance and chain parameters. Understanding circuit networks in this context paves the way for learning about S-parameters, which are more difficult to understand without a fundamental knowledge of circuit networks.

S-parameters are based on the character of signals propagating along a transmission line. The electrical properties of transmission lines are reviewed in connection with the fundamental theory of propagation. We will demonstrate how to use a special graphing technique using a Smith chart. In this context, we introduce S-parameters and tie them to the properties of electrical networks. Signal propagation, power flow, and circuit response are derived using S-parameters. Common transmission line construction techniques are presented with an emphasis on the microstrip transmission line.

Circuit analysis and synthesis are complementary processes. Analysis reveals the properties of existing circuits, and synthesis creates circuits from a list of requirements. We will show how to analyze high-frequency circuits using the new tools of S-parameters and the Smith chart. We will also introduce mapping functions and signal flow graphs for analyzing circuits. Mapping functions assist in the analysis of networks for which a particular performance, such as stability or constant gain, is desired, Signal flow graphs are useful for analyzing complicated circuits or interconnected networks. Synthesizing RF and microwave circuits focuses primarily on impedance matching. Several impedance-matching design techniques are shown along with their applications. These include lumped-element and transmission line circuits.

Designing RF and microwave amplifiers makes use of many of the techniques presented in this book. We will begin by describing the information available in transistor data sheets. We then explain circuit stability and present analytical methods of stability analysis using S-parameters. Amplifiers are designed by determining the desired impedance or reflection coefficient of the input and output circuits of the transistor. The transistor is matched to the input and output transmission line by these circuits. We will discuss methods of determining the reflection coefficient of these matching circuits for narrowband applications. This depends upon the desired gain and VSWR of the amplifier. A brief overview of broadband matching techniques is also provided.

A discussion of high frequency amplifier design would not be complete without addressing noise and noise figure. We will discuss the theory of noise in electronic circuits at high frequencies and define noise figure and noise measure of a network. We will use these definitions to modify the amplifier design methods to use noise as well as gain as a design criterion. This results in a wide variety of matching choices for a particular transistor in which trade-offs between gain and noise are considered. To help in making these choices, we show how a system's noise figure depends on the performance of each element or amplifier. Finally, the impact of system noise on signal-to-noise ratio is described.

Introduction

This book provides an introduction to RF and microwave amplifier design. As with all areas of electrical engineering, special components, materials and skills are needed to design RF and microwave amplifiers. Covered in this book are techniques and examples for designing small signal amplifiers. The term small signal refers to the amplifier application, not necessarily the actual size of the signal. A small signal amplifier is a linear circuit in which the input or output signal strength does not affect the circuit's electrical properties. If we were to change the signal amplitude at the input to the amplifier, the output signal would change in the same proportions. For example, an increase in the input by 5 dB would result in an increase in the output by 5 dB. We use analysis and design techniques that are based in linear network theory.

At microwave frequencies, special components are used. We will use microwave transistors in these design examples. To build small signal amplifiers, high-frequency transistors, capacitors and resistors are used, which are specially designed for these frequency bands. This book focuses on the design of single-transistor amplifiers. These transistors are manufactured and sold for use in certain frequency bands. Their electrical properties have been optimized to perform a specific task such as power amplification or receiver front ends. In addition, special design considerations must be made at high frequencies.

As a signal frequency increases, waves "bounce" between components. They cease to be static and begin to interact. Signals propagate both in a wire and through components. We cannot join wires and components without considering how these propagating waves will be reflected from the junctions. The wavelength of these signals becomes comparable in size to the circuit itself. A voltage source at one end of a wire will look similar to a current source one-quarter of a wavelength away. At a half wavelength, the source will again appear to be a voltage source, but the polarity will be reversed. When we build high-frequency circuits, size and length become an important consideration. Separating two components by a few feet or even a few millimeters can invert signal voltages or turn a voltage signal into a current signal. Instead of treating this as a disadvantage, we can use a length of wire itself to act as a capacitor, or transform one impedance that may be difficult to use to another, more easily handled impedance.

Not only does circuit size become critical at high frequencies, but the properties of materials also change. There are components that are specially manufactured to be used at high frequencies. Transistors and capacitors make up the bulk of these special components. However, connectors, circuit board material, and power splitters also play an important role in circuit design. These products, and the companies who manufacture them, can be found in any RF or microwave journal publication. In addition, we also utilize lengths of transmission lines as circuit elements.

High-frequency amplifiers are also built on a semiconductor chip, commonly called monolithic microwave integrated circuits (MMIC). Amplifiers built on a monolithic substrate are much smaller than comparable circuits built from discrete components. Diodes, transistors, capacitors, inductors, and resistors are fabricated in miniature. The design techniques presented in this book show how to use these discrete components and transmission lines.

1.1 The RF and Microwave Spectrum

Microwave and millimeter-wave frequency bands are shown in Figure 1-1. The two bands are clearly defined to be 1-30 GHz for the microwave band and 30-300 GHz for the millimeter-wave band. Radio Frequencies (RF) extend very low in frequency; some RF communication systems operate under 100 Hz. We usually use high-frequency design techniques when designing amplifiers over 200 MHz. However, digital and analog integrated circuits are operating at ever higher frequencies, providing compact, off-the-shelf applications at frequencies exceeding a couple of gigahertz. Although full systems are much easier to design with these modern components, we still must pay attention to circuit board layout and assembly.

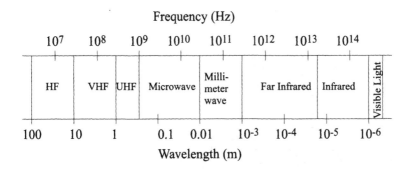

Figure 1-1 *The frequency spectrum showing different frequency bands.*

Microwave and millimeter wave bands are broken down further into smaller bands. Each band is referred to by a letter. Table 1-1 lists the bands and their numerical designation. Transistor manufacturers will sometimes refer to these band letters when describing transistors. For example, NEC describes its NE76184A transistor as "General Purpose L to Ku-Band MESFET" in their data sheets. The manufacturer describes this transistor as "appropriate for use in the second or third stages of low noise amplifiers operating in the 1-12 GHz frequency range" [1].

Table 1-1 *Microwave frequency band designations.*

Band	Frequency (GHz)
L-Band	1-2
S-Band	2-4
C-Band	4-8
X-Band	8-12
Ku-Band	12-18
K-Band	18-26.5
Ka-Band	26.5-40
Q-Band	40-60
W-Band	60-95

Other band designators break the frequency spectrum into decade bands. They have been used for many years and are shown in Table 1-2.

Table 1-2 *Universal frequency band designations.*

Band	Frequency
Very Low Frequency (VLF)	3 – 30 kHz
Low Frequency (LF)	30 – 300 kHz
Medium Frequency (MF)	300 kHz – 3 MHz
High Frequency (HF)	3 – 30 MHz
Very High Frequency (VHF)	30 – 300 MHz
Ultra High Frequency (UHF)	300 MHz – 3 GHz
Extra High Frequency (ELF)	3 – 30 GHz
Super High Frequency (SHF)	30 – 300 GHz

We can see from Figure 1-1 that signal wavelengths shorten in the microwave and millimeter-wave frequency regions. Instead of having simple voltages and currents that are constant between circuit components, we have waves that propagate between components. Junctions and discontinuities in the circuits cause reflections in these propagating waves. If a reflection is allowed to occur, the original wave and reflected wave will set up an interference pattern along the transmission path. These interference patterns are called *standing waves*. They are very similar to the wave patterns that can be seen on the surface of a pond when waves hit an object. To prevent unintentional reflections, the waves must be guided by transmission line structures from one place to the next. Common transmission lines are coaxial cables that carry television signals into the home and twisted-pair cables that carry network data between computers. We will show how to use transmission lines in different ways, from simple circuit interconnects to simulating capacitors and inductors at microwave frequencies.

1.2   Networks and Circuit Design

Chapters 2 through 4 focus on the mathematics of linear networks. Chapter 2 defines Z-, Y- and chain parameters and serves as an introduction to linear networks. Linear network theory provides the structured approach from which design methods are developed. First, we introduce networks which are characterized by certain parameters. A network is a circuit with well-defined terminals. It is treated as a "black box" in which current or voltage is imposed on one pair of terminals while affecting another terminal. How a network reacts to voltage or current excitation at its terminals is characterized by a set of parameters, or equations.

Chapter 3 explains transmission line theory and wave propagation and introduces the Smith chart. We look at wave reflection and standing waves on transmission lines and develop equations that are not intuitive or easy to visualize. The Smith chart gives us a graphical way of tackling problems. Many transmission line difficulties can be solved with a Smith chart, ruler, and pencil. Chapter 3 also introduces a new set of network parameters, called S-parameters, which are more suitable to high-frequency design. S-parameters describe a network in terms of power wave reflection and transmission from and through a circuit. As we study signal propagation in transmission lines, we can connect these transmission lines to circuits described by S-parameters. The performance of these systems can then be more easily analyzed.

Chapter 4 shows methods of analyzing microwave circuits using network S-parameters. We assume that we are given a circuit and need to determine the S-parameters. Our tools in the analysis of high-frequency circuits are network transformations and signal flow graphs. Once the foundations of network, transmission line, and S-parameter theory have been introduced, the methods of analyzing circuits are given and the actual circuit design process can begin.

1.3   Amplifier Stability, Gain and Noise Figure Design

In this book, we concentrate on the design of single-transistor amplifier circuits. Circuit discontinuities and components easily reflect RF and microwave signals. A considerable part of designing microwave amplifiers is synthesizing the input and output circuits. Chapter 5 introduces a few circuit synthesis techniques that can be used to design narrowband amplifiers. These circuits match the input and output of a transistor to transmission lines, antennas, and other circuits. Figure 1-2 shows a single-stage amplifier and its input and output matching circuits.

Chapter 6 describes how to design and characterize narrowband amplifiers. These amplifiers will consist of an input circuit, a transistor, and an output circuit. The transistor will be described by network parameters. It is important to study the amplifier's stability. There is not much isolation between the input and output of a high-frequency transistor. The input and output circuit affect each other, creating possible oscillation conditions in the circuit. This problem is comparable to a race condition in a digital circuit and must be avoided when designing microwave amplifiers. Once we are sure there is no potential instability in the transistor, we derive the matching conditions on the input and output so that a certain frequency can be obtained.

Chapter 7 introduces broadband amplifier design methods and provides a brief discussion of the design techniques. We show the importance of transistor modeling and how the model is used in synthesizing broadband matching circuits. The trade-off between gain and bandwidth, which depends on the quality factor (*Q*) of the transistor, is determined by the transistor model. We show how to design a broadband impedance transformer and describe matching by mimicking a filter and using lossy-tuned circuits.

Chapter 8 discusses noise and noise modeling with voltage sources, current sources, or power waves. Noise is created by thermal agitation of molecules and has a defined spectrum. Additional noise is created in active devices such as transistors and diodes. We can model a transistor as a noiseless device with imaginary voltage and/or current sources supplying noise to the circuit. The amount of excess noise that is picked up by a signal is called the noise figure of an amplifier. We provide the definition of noise figure and derive the relationship between the transistor and the rest of the circuit.

Finally, Chapter 9 shows how to design low noise amplifiers. Some transistor data sheets give the noise parameters of the device. These parameters are used, along with the S-parameters, to design an amplifier with a certain noise figure and gain. The trade-off between low noise figure and high gain is shown and ways of evaluating noise figure and the dynamic range of many cascaded microwave circuits are indicated.

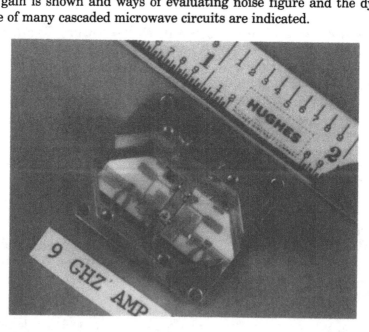

Figure 1-2 *Photograph of a 9 GHz single stage amplifier (courtesy of the Hughes Electronics Corporation).*

1.4   Microwave Systems

In this book, we assume that the reader has an engineer's basic under-standing of electronics and some experience with transmission lines and electromagnetics. The RF and microwave circuits mentioned here only apply to the high-frequency response of the circuit and the designer will still need to supply the correct bias currents and voltages. The designer also will need to know how to use capacitors to block bias currents, or bypass bias components to prevent inadvertent feedback.

The appendices at the end of some chapters list MatLab (by MathWorks) files that can be used to perform certain calculations. Many of the examples have been simulated with Eagleware's =SuperStar= circuit simulation soft-ware. Even the most careful design and modeling can result in a circuit that does not perform as expected, and small adjustments may be required. The circuit may have to be tuned by changing component values or the length or width of the transmission lines.

The techniques in this book provide a solid foundation for narrowband small signal amplifier design. Microwave systems, such as receivers and radar, require various other types of circuits. Power amplifiers usually operate in the large signal region. Amplifiers are the most common kind of circuit in microwave systems although components such as mixers, filters, switches, and oscillators are also needed. Since most microwave systems have more than one stage of frequency downconversion, there are many opportunities to design different amplifiers at different frequencies within one receiver or radar system.

1.5   References

1.   *NEC NE76184A General Purpose L to Ku-Band MESFET*, California Eastern Laboratories.

2

Introduction to Networks

2.1 Introduction

Microwave amplifiers are designed using mathematical techniques that require the use of network parameters. These network parameters describe how a circuit responds to certain input stimuli. Most amplifiers are designed by determining the circuit's function and then creating a circuit to accomplish something very close to what is required. For example, the circuit might be an amplifier with a specified gain and input and output impedance. This circuit is designed by forming an electronic network of various components. Using calculations, we can predict the circuit's gain and impedance match. To make these predictions, we must have a uniform method of describing circuits or networks of components and the circuit descriptions must be characterized mathematically. Circuit parameters are defined to represent the actions and responses of a circuit in mathematical terms. When we postulate a stimulus, e.g., when a television signal is being sent down a coaxial cable, we can calculate the likely circuit response to that signal.

This chapter introduces three types of circuit parameters describing electronic networks. These parameters, called the Z-, Y-, and chain parameters, are common in designing circuits up to two gigahertz but are not often used in microwave circuit design. We begin with these parameters because they are easy to understand and used in most computer-aided design (CAD) programs.

A circuit, or network, is defined by identifying pairs of terminals that will be used as an input or output. Often one of the terminals in a pair is a ground and may not be explicitly shown on a schematic. Each terminal pair is called a port. Each network has a certain number of these ports, and each port has two terminals. Signals are applied or measured at each of these ports. Current entering one terminal of a port must equal the current leaving the other terminal. This current and the voltage between terminals is used to derive the network parameters. The network parameters describe how the circuit reacts when signals are applied to any or all of the terminals of a network.

A circuit must have at least one port to be useful. Most circuits covered in this book will have two ports. For example, an amplifier will have an input port and an output port. Some circuits can have many ports depending on the circuit's function. We will develop the concept of network parameters by assuming any number of ports, or terminal pairs, and we will demonstrate how these parameters are found by using networks with one or two ports. A complete description of a network requires the calculation of a parameter or equation for each combination of input and output ports. A one-port network has one parameter; a two-port network has four parameters: two parameters for when Port 1 is used as the input, and two more for when Port 2 is used as the input.

Linear algebra is used with the stimulus and response forming a column matrix and the circuit parameters forming a square matrix to formalize these parameters. Employing these matrices enables us to borrow many fundamental principles from linear algebra. Finding a circuit's network parameters can be a lot of work, but once they are obtained, the design process will be considerably easier.

Section 2.2 introduces the open-circuit Z-parameters of a network. Z-parameters describe how a circuit responds to a current imposed on one port of the network. After the voltage at each terminal pair is found, the Z-parameters can be determined. The voltages are calculated by assuming that the port has a perfect open circuit connected to it. After defining Z-parameters, we then show how to find the Z-parameters of a network with one or two ports. Section 2.3 introduces the short-circuit Y-parameters, which describe how a circuit responds to a voltage source. This response is measured by shorting the two terminals of each port and finding the current flowing through the shorted terminals. We will illustrate the calculation of Y-parameters by finding the one-port and two-port Y-parameters of real circuits. Section 2.4 presents chain parameters, or ABCD-parameters, which use a combination of voltage and current excitation. Chain parameters have meaning only for two-port networks and describe how current and voltage imposed on port-one of a network appear as open-circuit voltage or short-circuit current on the output of the two-port.

2.2   Introduction to Z-Parameters

Z-parameters are used to describe how a circuit responds to current excitation on the terminal pairs of a network. Figure 2-1 shows a general n-port consisting of a circuit inside the box with two n terminals arranged in n pairs. We determine the Z-parameters by connecting a test current source to one terminal pair, loading the other terminal pairs with an infinite load resistance (a perfect open), and finding the voltage at each port and at each frequency of interest. The Z-parameters are the ratios of the open-circuit voltage and test current of each possible combination of test and measurement port configurations. This is a concept we will explain when we describe how this general formulation is applied to one- and two-port networks.

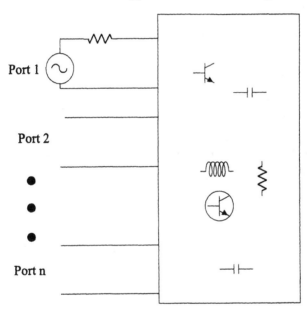

Figure 2-1 *An n-port network with a test source connected to Port 1 and load impedances connected to the other ports.*

One-port devices are the simplest networks with two terminals. Finding the Z-parameters for these networks is the same as finding the input impedance. A current test source is connected to the terminals, as shown in Figure 2-2. The current flowing in the top terminal must be equal to the current following out of the bottom terminal otherwise it cannot be considered a one-port device (there must be no current flowing in or out of the network through an unseen terminal). The one-port Z-parameter, Z_{11}, or input impedance, is found by determining the voltage V_1 between the two terminals at port-one due to the excitation current I_1.

$$V_1 = Z_{11}I_1 \qquad\qquad 2.1$$

We use two subscripts on these one-port parameters to conform to the conventional nomenclature for multi-ports.

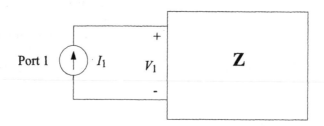

Figure 2-2 *A one-port with test source connected.*

Even though the description given for a one-port network may seem trivial, as the number of components "inside" the one-port increases, it can become a challenge to determine the input impedance of the network.

Example 2.1: Given the circuit in Figure 2-3, find the one-port Z-parameter of this network.

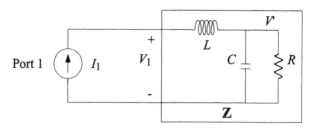

Figure 2-3 *Circuit for Examples 2.1 and 2.4.*

$$Z_{11} = \frac{V_1}{I_1}$$

$$V' = I_1 \left(\frac{R \, \dfrac{1}{sC}}{R + \dfrac{1}{sC}} \right)$$

$$V_1 = sLI_1 + I_1\left(\frac{R}{sCR+1}\right)$$

$$Z_{11} = sL + \left(\frac{R}{sCR+1}\right)$$

The Z-parameters of two-port networks are formulated in the following way. A generalized four-terminal network with two pairs of terminals is shown in Figure 2-4. This network is considered a two-port *only if* the four currents leaving its terminals satisfy

$$I_1 = I_1' \text{ and } I_2 = I_2'$$ 2.2

The nomenclature we will use is a capital Z with two subscripts when describing Z-parameters. Because these parameters are values of impedance, the impedance is the value of Z_{ij} needed to satisfy

$$V_i = Z_{ij}I_j$$

where V_i is the voltage at port i and I_j is the current at port j.

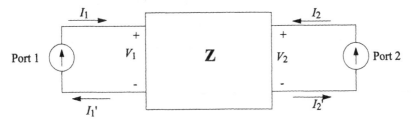

Figure 2-4 *A two-port network.*

The first subscript is the port number where the open-circuit voltage is measured, and the second subscript is the port number where the test current source is exiting the network. Figure 2-5(a) shows a two-port network with an ideal current source connected to Port 1, Figure 2-5(b) shows a two-port with a current source connected to Port 2. The Z-parameters of the network are given by

$$V_1 = Z_{11}I_1, \quad V_2 = Z_{21}I_1$$ 2.3

where V_1 and V_2 are shown on the circuit in Figure 2-5(a) with the current source connected to Port 1 and

$$V_1 = Z_{12}I_2, \quad V_2 = Z_{22}I_2 \tag{2.4}$$

where V_1 and V_2 are shown on the circuit in Figure 2-5(b). By superposition, the total voltage at Port 1 due to a current source on any or all ports of the network is

$$V_1 = Z_{11}I_1 + Z_{12}I_2 \tag{2.5}$$

The total voltage at Port 2 is similarly found to be

$$V_2 = Z_{21}I_1 + Z_{22}I_2 \tag{2.6}$$

The parameters Z_{11}, Z_{21}, Z_{12}, and Z_{22} depend only on the internal circuit components and how they are connected. The two equations above constitute a set of linear equations and can be written in a matrix format by combining the two voltages and two currents into column vectors. The Z-parameters form a two-by-two matrix.

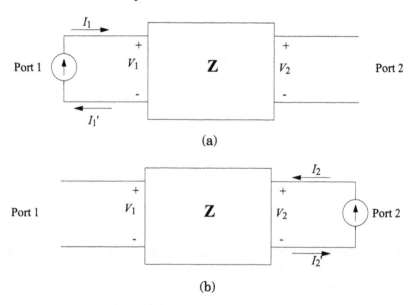

(a)

(b)

Figure 2-5 *A two-port network with (a) a test source connected to Port 1 and (b) a test source connected to Port 2.*

$$\begin{bmatrix} V_1 \\ V_2 \end{bmatrix} = \begin{bmatrix} Z_{11} & Z_{12} \\ Z_{21} & Z_{22} \end{bmatrix} \begin{bmatrix} I_1 \\ I_2 \end{bmatrix} \tag{2.7}$$

or

$$\mathbf{V} = \mathbf{ZI}$$ 2.8

where $\mathbf{Z}$ is called the open-circuit impedance matrix. The elements of this matrix are the Z-parameters of the network.

Example 2.2: Find the two-port Z-parameters of the circuit shown in Figure 2-6.

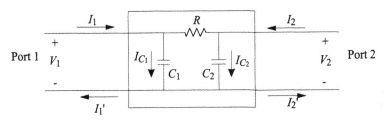

Figure 2-6 *Circuit for Example 2.2 and Example 2.5.*

The law of superposition allows us to break up the two-port into two test circuits, as shown in Figure 2-5. We use the test circuit of Figure 2-5(a) to find Z_{11} and Z_{21} because the current (I_2) equals zero. Equation 2.5 shows that V_1 is only a function of I_1 and Z_1.

$$Z_{11} = \frac{V_1}{I_1}\bigg|_{I_2=0}$$

$$I_1 = V_1 s C_1 + \frac{V_1}{\left(R + \dfrac{1}{s\,C_2}\right)}$$

$$= V_1 s C_1 + \frac{V_1 s C_2}{R s C_2 + 1}$$

$$= \frac{V_1\left(s C_2 + s^2 C_1 C_2 + s C_2\right)}{R s C_2 + 1}$$

$$Z_{11} = \frac{s\,C_2\,R + 1}{s^2 C_1 C_2 + s C_2 + 1}$$

The expression for Z_{21} is derived from Equation 2.6.

$$Z_{21} = \frac{V_2}{I_1}\bigg|_{I_2 = 0}$$

We find V_1 and V_2 as a function of current flowing through capacitor C_2.

$$V_1 = \left(\frac{1}{sC_2} + R\right)I_{C_2}$$

$$V_2 = I_{C_2}\frac{1}{sC_2}$$

Then V_2 can be expressed as a function of V_1,

$$V_2 = \frac{\dfrac{1}{s\,C_2}}{R + \dfrac{1}{s\,C_2}}V_1 = \frac{1}{sRC_2 + 1}V_1$$

and we can use Z_{11} to relate these two voltages to the current flowing into Port 1.

$$V_1 = I_1 Z_{11} = I_1\frac{s\,C_2\,R + 1}{s^2 RC_1 C_2 + s\,(C_1 + C_2)}$$

$$V_2 = I_1\left(\frac{s\,C_2\,R + 1}{s^2 RC_1 C_2 + s\,(C_1 + C_2)}\right)\frac{1}{sC_2 R + 1}$$

$$Z_{21} = \frac{1}{s^2 RC_1 C_2 + s\,(C_1 + C_2)}$$

The test circuit of Figure 2-5(b) is used to find Z_{22} and Z_{12}. The input current I_1 is equal to zero, and Z_{22} is found by relating I_2 to V_2.

$$Z_{22} = \frac{V_2}{I_2}\bigg|_{I_1 = 0}$$

$$I_2 = V_2 sC_2 + \frac{V_2}{\left(R + \dfrac{1}{sC_1}\right)}$$

$$Z_{22} = \frac{s\,C_1\,R+1}{s^2 RC_1 C_2 + s\,(C_1 + C_2)}$$

The expression for Z_{12} comes from Equation 2.5, where I_1 equals zero.

$$Z_{12} = \frac{V_1}{I_2}\bigg|_{I_1=0}$$

We find the current flowing through capacitor C_1 as a function of both V_1 and V_2.

$$V_2 = \left(\frac{1}{sC_1} + R\right) I_{C_1}$$

$$V_1 = I_{C_1} \frac{1}{sC_1}$$

$$V_1 = \frac{\dfrac{1}{s\,C_1}}{R + \dfrac{1}{s\,C_1}} V_2 = \frac{1}{sRC_1 + 1} V_2$$

We use Z_{22} to relate I_2 to V_2

$$V_2 = I_2 Z_{22} = I_2 \frac{s\,C_1\,R+1}{s^2 RC_1 C_2 + s\,(C_1 + C_2)}$$

and find Z_{12}.

$$V_1 = I_2 \left(\frac{s\,C_1\,R+1}{s^2 RC_1 C_2 + s\,(C_1 + C_2)}\right) \frac{1}{sC_1 R + 1}$$

$$Z_{12} = \frac{1}{s^2 RC_1 C_2 + s\,(C_1 + C_2)}$$

The open-circuit Z-parameters of the network in Figure 2-6 are

$$\mathbf{Z} = \begin{bmatrix} \dfrac{sRC_2+1}{s^2RC_1C_2+s\,(C_1+C_2)} & \dfrac{1}{s^2RC_1C_2+s\,(C_1+C_2)} \\[3ex] \dfrac{1}{s^2RC_1C_2+s\,(C_1+C_2)} & \dfrac{sRC_1+1}{s^2RC_1C_2+s\,(C_1+C_2)} \end{bmatrix}$$

Some networks cannot be expressed with Z-parameters because ideal current sources are part of their definition. Consider the network shown in Figure 2-7. When an ideal current source is connected to Port 1, there is no current flowing into or out of either port since there is no complete path for current to flow. Also, recall that the voltage across an ideal current source is not defined unless there is a shunt impedance to develop a voltage. We need other parameters to describe networks such as the one shown in Figure 2-7.

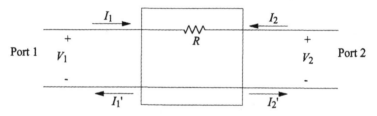

Figure 2-7 *A simple network where the Z-parameters are undefined.*

Z-parameters are ideally suited for describing networks that are connected in series (see Figure 2-8). The network in this figure shows two two-ports connected in series. Remember that the definition of a two-port requires that

$$I_1 = I_1', \ I_1 = I_1', \ I_2 = I_2', \text{ and } I_2 = I_2' \qquad 2.9$$

Furthermore, it is necessary that

$$I_1' = I_1^\dagger \text{ and } I_2' = I_2^\dagger \qquad 2.10$$

in order for the current vectors I_1 and I_2 to be the same in and out of each port. The voltages at the terminals of the combined network are equal to the sum of the voltages at the individual ports.

$$V_1^T = V_1 + V_1^\dagger \text{ and } V_2^T = V_2 + V_2^\dagger \qquad 2.11$$

Using vector notation, the Z-parameters of the two two-ports connected in series can be easily found.

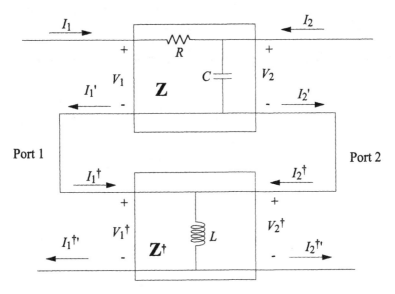

Figure 2-8 *Two two-port networks connected in series.*

$$\mathbf{V^T} = \mathbf{V} + \mathbf{V^\dagger} \qquad\qquad 2.12$$

$$\mathbf{V^T} = \mathbf{ZI} + \mathbf{Z^\dagger I^\dagger} \qquad\qquad 2.13$$

Since

$$\mathbf{I} = \mathbf{I^\dagger}$$

the series connection of both networks is

$$\mathbf{V^T} = \left(\mathbf{Z} + \mathbf{Z^\dagger}\right)\mathbf{I} \qquad\qquad 2.14$$

The Z-parameters of series-connected two-ports are found by simply adding the Z-parameters of the individual two-ports.

$$\begin{bmatrix} Z_{11}{}^T & Z_{12}{}^T \\ Z_{21}{}^T & Z_{22}{}^T \end{bmatrix} = \begin{bmatrix} Z_{11} & Z_{12} \\ Z_{21} & Z_{22} \end{bmatrix} + \begin{bmatrix} Z_{11}{}^\dagger & Z_{12}{}^\dagger \\ Z_{21}{}^\dagger & Z_{22}{}^\dagger \end{bmatrix} \qquad 2.15$$

Networks of more than two ports are analyzed the same way as the two-ports above. A test current source is connected to each terminal pair, and the voltage appearing at all the terminal pairs is calculated. Z_{nm} is found by dividing the voltage at port n by the test current at port m when port n is open. At times, we need to look at networks of more than two ports, and networks of three or more ports are common in more complex circuits and network analyses of large circuits.

Example 2.3: Find the two-port Z-parameters of the two series-connected two-port networks shown in Figure 2-9.

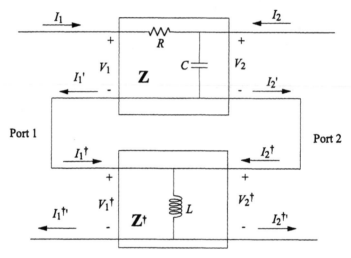

Figure 2-9 *Circuit for Example 2.3.*

First, we determine the Z-parameters for the top two-port.

$$Z_{11} = \left.\frac{V_1}{I_1}\right|_{I_2 = 0}$$

$$I_1 = \frac{s\,C\,V_1}{(s\,C\,R+1)}$$

$$Z_{11} = \frac{(s\,C\,R+1)}{s\,C}$$

$$Z_{21} = \left.\frac{V_2}{I_1}\right|_{I_2 = 0}$$

$$V_2 = \frac{V_1}{(s\,C\,R+1)}$$

$$Z_{21} = \frac{1}{sC}$$

$$Z_{12} = \frac{V_1}{I_2}\bigg|_{I_1 = 0}$$

$$Z_{12} = \frac{1}{sC}$$

$$Z_{22} = \frac{V_2}{I_2}\bigg|_{I_1 = 0}$$

$$Z_{22} = \frac{1}{sC}$$

$$\mathbf{Z} = \begin{bmatrix} R + \dfrac{1}{sC} & \dfrac{1}{sC} \\ \dfrac{1}{sC} & \dfrac{1}{sC} \end{bmatrix}$$

Next, we find the Z-parameters for the bottom two-port. All elements of the $\mathbf{Z}^\dagger$ matrix are equal to sL.

$$\mathbf{Z} = \begin{bmatrix} sL & sL \\ sL & sL \end{bmatrix}$$

The Z-parameters of the combined network are the sum of the two Z-parameter matrices.

$$\mathbf{Z}^{\mathbf{T}} = \begin{bmatrix} \dfrac{sC(sL+R)+1}{sC} & \dfrac{s^2CL+1}{sC} \\ \dfrac{s^2CL+1}{sC} & \dfrac{s^2CL+1}{sC} \end{bmatrix}$$

2.3 Introduction to Y-Parameters

Y-parameters are used to describe a circuit response to an ideal voltage source at the terminal pairs of a network. The Y-parameters of the network in Figure 2-1 are found by connecting an ideal voltage source to one terminal pair, shorting the other terminal pairs together, and finding the current flowing through these shorts on each port. The Y-parameter is the ratio of this current and the test voltage.

Finding the Y-parameters for one-port networks is the same as finding the input admittance. When a voltage test source is connected to the terminals, as shown in Figure 2-10, we can solve for the current flowing into the port. The one-port Y-parameter (Y_{11}), is found by solving the following equation for Y_{11}.

$$I_1 = Y_{11}V_1 \qquad\qquad 2.16$$

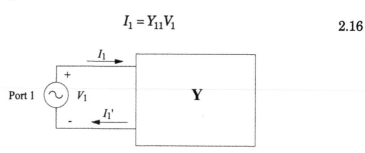

Figure 2-10 *A one-port network with a test source connected.*

Example 2.4: Given the circuit in Figure 2-3, find the one-port Y-parameter of the network.

$$Y_{11} = \frac{I_1}{V_1}$$

$$I_1 = \frac{s\,C\,V_1}{(s\,C\,R+1)}$$

$$Y_{11} = \frac{s^2CLR+sL+R}{sCR+1}$$

The Y-parameters and Z-parameters of two-port networks are formulated in a similar way. Figure 2-11(a) shows the two-port network with an ideal voltage source connected to Port 1, and Figure 2-11(b) shows the two-port network with a voltage source connected to Port 2. The Y-parameters of the network are given by

$$I_1 = Y_{11}V_1, \quad I_2 = Y_{21}V_1 \qquad\qquad 2.17$$

where I_1 and I_2 are shown on the circuit in Figure 2-11(a) with the voltage source connected to Port 1 and

$$I_1 = Y_{12}V_2, \quad I_2 = Y_{22}V_2 \qquad\qquad 2.18$$

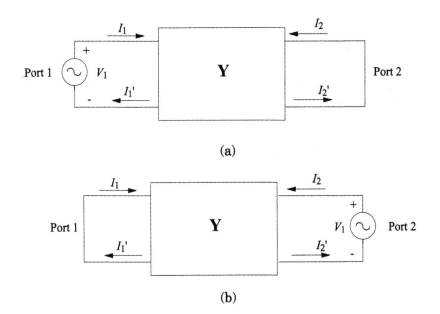

(a)

(b)

Figure 2-11 *A two-port with (a) a test source connected to Port 1 and (b) a test source connected to Port 2.*

where I_1 and I_2 are shown on the circuit in Figure 2-11(b). By superposition, the total current at Port 1 due to a voltage source on any or all ports of the network is

$$I_1 = Y_{11}V_1 + Y_{12}V_2 \qquad\qquad 2.19$$

The total current at Port 2 is similarly found to be

$$I_2 = Y_{21}V_1 + Y_{22}V_2 \qquad\qquad 2.20$$

The parameters Y_{11}, Y_{21}, Y_{12}, and Y_{22} depend only on the internal circuit components and how they are connected. These Y-parameters form a two-by-two matrix.

$$\begin{bmatrix} I_1 \\ I_2 \end{bmatrix} = \begin{bmatrix} Y_{11} & Y_{12} \\ Y_{21} & Y_{22} \end{bmatrix} \begin{bmatrix} V_1 \\ V_2 \end{bmatrix} \qquad\qquad 2.21$$

$$\mathbf{I} = \mathbf{YV} \qquad\qquad 2.22$$

Y is called the short-circuit admittance matrix and the elements of this matrix are the Y-parameters of the network.

Example 2.5: Find the two-port Y-parameters of the circuit shown in Figure 2-6. The test circuit in Figure 2-11(a) shows that the voltage at Port 2 is equal to zero. Under this condition in Equation 2.19, Y_{11} is given in terms of I_1 and V_1.

$$Y_{11} = \frac{I_1}{V_1}\bigg|_{V_2 = 0}$$

$$I_1 = \frac{V_1}{R} + sC_1V_1$$

$$Y_{11} = \frac{sC_1R + 1}{R}$$

Y_{21} can be found by using Equation 2.20 and Figure 2-11(a).

$$Y_{21} = \frac{I_2}{V_2}\bigg|_{V_2 = 0}$$

$$I_2 = -\frac{V_1}{R}$$

$$Y_{21} = -\frac{1}{R}$$

Y_{22} is found by using the test circuit shown in Figure 2-11(b) and Equation 2.20.

$$Y_{22} = \frac{I_2}{V_2}\bigg|_{V_1 = 0}$$

$$I_2 = \frac{V_2}{R} + V_2sC_2$$

$$Y_{22} = \frac{sC_2R+1}{R}$$

Using Equation 2.19 and Figure 2-11(b), we can find Y_{12}.

$$Y_{12} = \frac{I_1}{V_2}\bigg|_{V_1 = 0}$$

$$I_1 = -\frac{V_2}{R}$$

$$Y_{12} = -\frac{1}{R}$$

The Y-parameters of the network in Figure 2-6 are

$$\mathbf{Y} = \begin{bmatrix} \dfrac{sC_1R+1}{R} & -\dfrac{1}{R} \\ -\dfrac{1}{R} & \dfrac{sC_2R+1}{R} \end{bmatrix}$$

Similar to the Z-parameters, some networks cannot be expressed with Y-parameters. Consider the network shown in Figure 2-12. When Port 2 is short-circuited, the voltage at this port is equal to zero, which makes the Y-parameters undefined for this network.

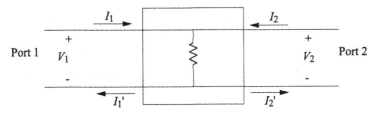

Figure 2-12 *A simple network where the Y-parameters are undefined.*

We can see that the Y-parameters of a network are related to the Z-parameters by multiplying both sides of Equation 2.22 by the inverse of **Y**.

$$\mathbf{Y}^{-1}\mathbf{I} = \mathbf{Y}^{-1}\mathbf{Y}\mathbf{V} \tag{2.23}$$

$$\mathbf{Y}^{-1}\mathbf{I} = \mathbf{V} = \mathbf{Z}\mathbf{I} \tag{2.24}$$

The Y-parameter matrix is the inverse of the Z-parameter matrix.

$$\mathbf{Y}^{-1} = \mathbf{Z} \tag{2.25}$$

Y-parameters are ideal for analyzing networks that are connected in parallel, as shown in Figure 2-13. The voltage at ports one and two of both two-ports is equal.

$$V_1^T = V_1 = V_1 \text{ and } V_2^T = V_2 = V_2 \qquad 2.26$$

Using Kirchhoff's current law, the current entering the composite network is equal to the sum of the current entering both two-ports.

$$I_1^T = I_1 + I_1^\dagger \text{ and } I_2^T = I_2 + I_2^\dagger \qquad 2.27$$

Using vector notation, the Y-parameters of the two two-ports connected in parallel is

$$\mathbf{I^T = I + I} \qquad 2.28$$

$$\mathbf{I^T = YV + Y\ V} \qquad 2.29$$

Since

$$\mathbf{V = V^\dagger}$$

the current vector at the network ports is

$$\mathbf{I^T = \left(Y + Y^\dagger\right)V} \qquad 2.30$$

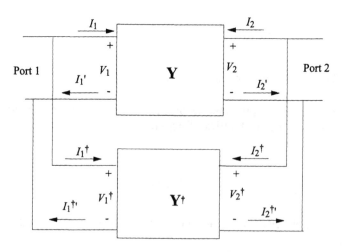

Figure 2-13 *Two two-ports connected in parallel.*

The Y-parameters of two two-ports connected in parallel are found by adding the two Y-parameter matrices.

$$\begin{bmatrix} Y_{11}^{T} & Y_{12}^{T} \\ Y_{21}^{T} & Y_{22}^{T} \end{bmatrix} = \begin{bmatrix} Y_{11} & Y_{12} \\ Y_{21} & Y_{22} \end{bmatrix} + \begin{bmatrix} Y_{11} & Y_{12} \\ Y_{21} & Y_{22} \end{bmatrix} \qquad 2.31$$

Example 2.6: Find the Y-parameters of the network shown in Figure 2-14. First, we need to determine the Y-parameter matrix for the top network. Y_{11} and Y_{21} are found using the test circuit of Figure 2-11(a), as applied to network **Y**.

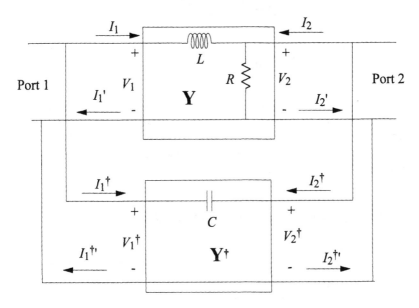

Figure 2-14 *Circuit for Example 2.6.*

$$Y_{11} = \frac{1}{sL}$$

$$Y_{21} = -\frac{1}{sL}$$

Y_{22} and Y_{12} of network Y are found using the test circuit of Figure 2-11(b).

$$Y_{22} = \frac{R + sL}{sLR}$$

$$Y_{12} = -\frac{R + sL}{sLR}$$

Next, we find the Y-parameters of the bottom network $\mathbf{Y}^\dagger$.

$$\mathbf{Y} = \begin{bmatrix} sC & -sC \\ -sC & sC \end{bmatrix}$$

Finally, we add the $\mathbf{Y}$ and $\mathbf{Y}^\dagger$ matrices together.

$$\mathbf{Y}^T = \begin{bmatrix} \dfrac{s^2CL+1}{sL} & -\dfrac{s^2CLR+sL+R}{sLR} \\ -\dfrac{s^2CL+1}{sL} & \dfrac{s^2CLR+sL+R}{sLR} \end{bmatrix}$$

We can extend this analysis to any number of ports and convert from Z- to Y-parameters as long as the matrices can be inverted. Complex networks can be analyzed by breaking the network into smaller circuits, calculating the Z- or Y-parameters of these smaller units, and then finding the parameters of the entire network.

2.4   Introduction to Chain Parameters

Chain, or ABCD-parameters are another common set of parameters used to describe networks. Chain parameters are ideal for analyzing two-port networks that are, or will be, connected in cascade (see Figure 2-15). Chain parameters are formulated for the analysis of cascaded two-ports and therefore apply only to two-port networks. As shown in Figure 2-15, the voltage at Port 2 of the first two-port is equal to the voltage at Port 1 of the second two-port. Furthermore, the current entering Port 2 of the first two-port is equal to the current leaving Port 1 of the second two-port. It is useful to have a set of network parameters where the voltage and current at any one port is in the same column vector. By rearranging the Z-parameter linear equations, we can solve for V_1 and I_1

$$-Z_{21}I_1 = Z_{22}I_2 - V_2 \qquad\qquad 2.32$$

which yields

$$I_1 = \frac{1}{Z_{21}} V_2 - \frac{Z_{22}}{Z_{21}} I_2 \qquad \text{2.33}$$

and

$$V_1 = Z_{11} I_1 + Z_{12} I_2 \qquad \text{2.34}$$

Substituting Equation 2.33 into 2.34, we receive an expression for the voltage at Port 1 as a function of the voltage and current at Port 2.

$$V_1 = \frac{Z_{11}}{Z_{21}} V_2 - \left(\frac{Z_{11} Z_{22}}{Z_{21}} - Z_{12} \right) I_2 \qquad \text{2.35}$$

Equation 2.35 and 2.33 are used to define the chain parameters

$$V_1 = AV_2 - BI_2 \qquad \text{2.36}$$

$$I_1 = CV_2 - DI_2 \qquad \text{2.37}$$

where

$$A = \frac{Z_{11}}{Z_{21}}$$

$$B = \frac{Z_{11} Z_{22}}{Z_{21}} - Z_{12}$$

$$C = \frac{1}{Z_{21}}$$

$$D = \frac{Z_{22}}{Z_{21}}$$

A, B, C, and D are the chain parameters. Their matrix form is

$$\mathbf{T} = \begin{bmatrix} A & B \\ C & D \end{bmatrix} \qquad \text{2.38}$$

Thus, the analysis of cascaded two-ports can be simplified. Consider the network shown in Figure 2-15, which has two two-ports connected in a cascade configuration. The voltage at the output of the first two-port is equal to the voltage at the input of the second two-port.

$$V_2 = V_1^\dagger \tag{2.39}$$

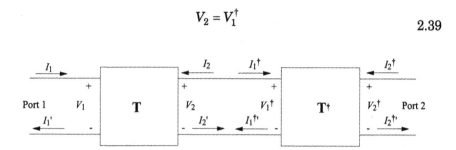

Figure 2-15 *Two two-port networks connected in cascade.*

Furthermore, current leaving the first two-port is equal to the current entering the second two-port, or in conventional network current directions,

$$I_2 = -I_1^\dagger \tag{2.40}$$

The chain parameters of the composite network are found to be

$$\begin{bmatrix} V_1 \\ I_1 \end{bmatrix} = \begin{bmatrix} A & B \\ C & D \end{bmatrix} \begin{bmatrix} V_2 \\ -I_2 \end{bmatrix} \quad \begin{bmatrix} V_1^\dagger \\ I_1^\dagger \end{bmatrix} = \begin{bmatrix} A^\dagger & B^\dagger \\ C^\dagger & D^\dagger \end{bmatrix} \begin{bmatrix} V_2^\dagger \\ -I_2^\dagger \end{bmatrix} \tag{2.41}$$

since

$$\begin{bmatrix} V_2 \\ -I_2 \end{bmatrix} = \begin{bmatrix} V_1^\dagger \\ I_1^\dagger \end{bmatrix}$$

$$\begin{bmatrix} V_1 \\ I_1 \end{bmatrix} = \mathbf{TT}^\dagger \begin{bmatrix} V_2^\dagger \\ -I_2^\dagger \end{bmatrix} \tag{2.42}$$

Determining the chain parameters of a two-port requires the use of both current and voltage test sources to be placed on Port 1 (the input port). Shorts and opens are needed on Port 2 (the output port), as shown in Figure 2-16.

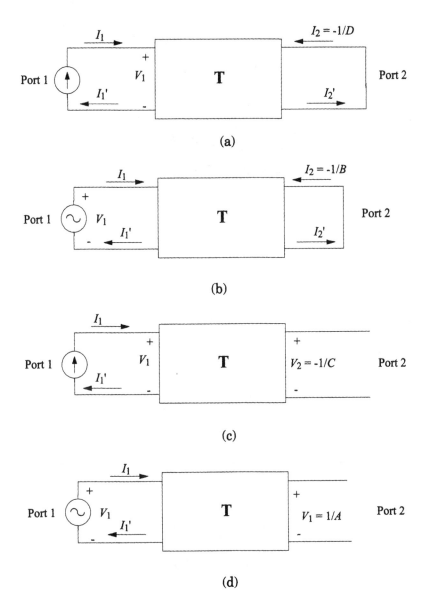

Figure 2-16 *The method for finding the chain parameters of a two-port.*

Example 2.7: Find the chain parameters of the network shown in Figure 2-17. First, the test circuit of Figure 2-16(d) is used to find A.

$$V_2 = V_1 \frac{\dfrac{1}{sC_2}}{sL + R + \dfrac{1}{sC_2}}$$

$$= V_1 \frac{1}{s^2 LC_2 + sC_2 R + 1}$$

$$A = s^2 LC_2 + sC_2 R + 1$$

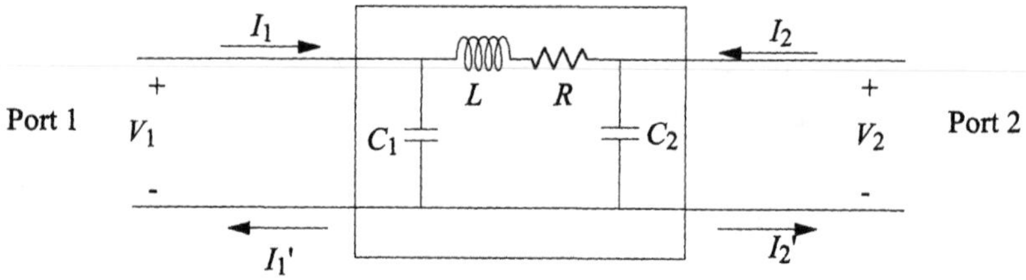

Figure 2-17 *The circuit for Example 2.7.*

Next, the test circuit shown in Figure 2-16(b) is used to find B.

$$I_2 = -V_1 \frac{1}{R + sL} = -\frac{1}{B}$$

$$B = R + sL$$

Then, the circuit in Figure 2-16(c) is used to find C. The voltage at the terminals of Port 1 is equal to V_1'.

$$V_1' = I_1 \frac{\dfrac{1}{sC_1}\left(R + sL + \dfrac{1}{sC_2}\right)}{\dfrac{1}{sC_1} + \dfrac{1}{sC_2} + R + sL}$$

$$V_2 = V_2' \frac{\dfrac{1}{sC_2}}{R + sL + \dfrac{1}{sC_2}}$$

$$V_2 = I_1 \frac{R + sL + \dfrac{1}{sC_2}}{sC_2 + sC_1 + s^2 RC_1 C_2 + s^3 LC_1 C_2}$$

$$C = \frac{s^2 LC_1 + sRC_1 + sC_1 + 1}{s^2 LC_2 + sRC_2 + 1}$$

Finally, the circuit in Figure 2-16(a) is used to find D. Again, we find the voltage at Port 1 (V_1).

$$V_1 = I_1 \frac{\dfrac{1}{sC_1}(R + L)}{\dfrac{1}{sC_1} + R + sL}$$

$$I_2 = V_1 \left(-\frac{1}{R + sL} \right)$$

$$I_2 = I_1 \frac{-1}{1 + sRC_1 + s^2 LC_1} = -\frac{1}{D}$$

$$D = 1 + sRC_1 + s^2 LC_1$$

In most cases, we can convert between Z-, Y-, and chain parameters. Chain parameters can be derived from the Z-parameters as described in Equation 2.37. Use the following equations to convert from Y-parameters to chain parameters:

$$A = -\frac{Y_{22}}{Y_{21}} \qquad\qquad 2.43$$

$$B = -\frac{1}{Y_{21}} \qquad\qquad 2.44$$

$$C = -\frac{Y_{11} Y_{22} - Y_{21} Y_{12}}{Y_{21}} \qquad\qquad 2.45$$

$$D = -\frac{Y_{11}}{Y_{21}} \qquad\qquad 2.46$$

In conclusion, converting to Z- and Y-parameters to form chain parameters is accomplished using these formulas:

$$\mathbf{Z} = \begin{bmatrix} \dfrac{A}{C} & \dfrac{\Delta_T}{C} \\ \dfrac{1}{C} & \dfrac{D}{C} \end{bmatrix} \qquad\qquad 2.47$$

$$\mathbf{Y} = \begin{bmatrix} \dfrac{D}{B} & \dfrac{-\Delta_T}{B} \\ \dfrac{-1}{B} & \dfrac{A}{B} \end{bmatrix} \qquad\qquad 2.48$$

where $\Delta_T = AD - BC$.

2.5 Summary

We have provided an introduction to Z-, Y-, and chain parameters. Their usage in the analysis and design of filters, impedance-matching networks, and RF circuits is discussed in detail in two excellent texts listed in the reference section at the end of this chapter. Understanding Z-, Y-, and chain parameters is paramount to comprehending concepts covered in the remainder of this book.

Z-parameters describe how a circuit or network responds to a current source when the network terminals are open. They are especially useful when two or more networks are connected in series. Y-parameters describe how a circuit responds to a voltage excitation by shorting the terminal pairs and calculating the current flowing through these shorts. When networks are connected in parallel, the Y-parameters of the composite network can be easily found by adding the Y-parameters of the individual networks. Chain parameters, or ABCD-parameters, are a convenient means to calculate what circuits do when they are connected in cascade. We have also shown how to convert from one of these three parameters to another. Large networks can be analyzed by breaking parameters into smaller networks connected in either parallel or cascade series.

The next chapter introduces the circuit parameters that are most useful in microwave design. Some data books, both old and new, describe transistors with Z- or Y-parameters. At microwave frequencies, most data books and design tutorials use S-parameters.

2.6   Problems

2.1   Find the Z-parameters of the one-port shown in Figure 2-18.

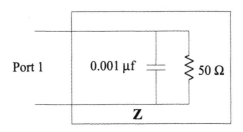

Figure 2-18 *Circuit for Problem 2.1.*

2.2   Find the Z-parameters of the two-port shown in Figure 2-19.

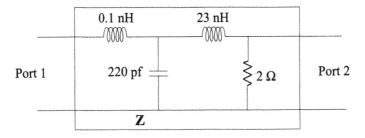

Figure 2-19 *Circuit for Problem 2.2.*

2.3   Find the Y-parameters of the one-port in Figure 2-20.

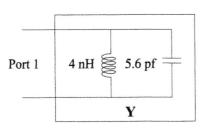

Figure 2-20 *Circuit for Problem 2.3.*

2.4   Find the Y-parameters of the two-port shown in Figure 2-21.

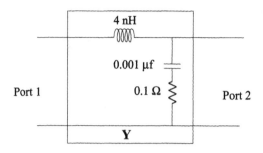

Figure 2-21 *Circuit for Problem 2.4.*

2.5 Invert the Z-parameters from Example 2.3 to obtain the Y-parameters.

2.6 Find the chain parameters of the two-port shown in Figure 2-22.

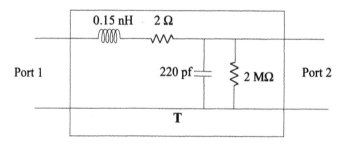

Figure 2-22 *Circuit for Problem 2.6.*

2.7 Derive the Y- to chain parameter equations.

2.8 Derive the equations that convert chain parameters to Z- and Y-parameters.

2.7 References

1. Mathaei, George L., Leo Young, and E. M. T. Jones. *Microwave Filters, Impedance-Matching Networks, and Coupling Structures.* New York: McGraw-Hill, 1964.

2. Temes, G. C. and J. W. LaPatra. *Circuit Synthesis and Design.* New York: McGraw-Hill, 1977.

3

Introduction to S-Parameters

3.1 Introduction

This chapter describes circuit parameters that are based on power flow. The open-circuit Z-parameters and the short-circuit Y-parameters are used in many of the circuit computer-aided design (CAD) programs. We can calculate the network parameters of circuits made from well-known components such as resistors, capacitors, and inductors. However, the circuit parameters of transistors and other types of semiconductors must be measured. As frequencies rise, it becomes increasingly difficult to measure voltage and current on the ports of a circuit.

This chapter introduces transmission lines and the mathematical expression that is used to describe their voltage and current. A detailed discussion of the electromagnetic foundation of transmission line theory is beyond the scope of this book. Instead, we focus on the propagation of power and signals and show how the power flow relates to voltage and current at any point along a transmission line. Transmission lines are an integral part of the transmission of power and signals in high-frequency circuits. In microwave circuits, rather than using resistors, capacitors, or inductors, transmission lines are employed to connect components and produce a complex impedance.

Further, we will develop another set of circuit parameters based on power flow in a transmission line. Whereas voltage and current are difficult to measure at microwave frequencies, power flow can be measured easily with directional couplers. Scattering parameters, or S-parameters, are ratios of power flow amplitude and phase in a circuit. S-parameters are usually listed in the transistor's data sheets. Since S-parameter power vectors are related to voltage and current in a transmission line, we will show the relation between S-, Z-, Y-, and chain parameters.

In the last section of this chapter, some of the more common transmission line technologies are introduced. Coaxial cables and waveguides are very common interconnected transmission media found in RF and microwave systems. Planar transmission lines are used in microwave circuits because these lines can easily be manufactured on printed circuit boards. Microwave circuits are commonly made using a microstrip transmission line structure. Such a structure consists of a conducting strip of a specified width and thickness suspended above a uniform ground plane by an insulating substrate or circuit board material. Microstrip lines are an efficient transmission medium and easy to manufacture. Elements of microstrip lines are used to interconnect components and create the complex impedances needed to design a circuit.

3.2   Transmission Lines

A transmission line is a generic term describing a medium or system of propagating energy from one point to another. As frequencies increase, the wires and interconnects between circuit components become larger compared to the wavelength of the signal flowing along them. Consider an interconnect between several components shown in Figure 3-1. At low frequencies, this wire can be considered part of the node connecting these components. At high frequencies, the wire's physical length introduces a noticeable delay to the signal traveling from one end to the other. For example, a 10 GHz signal has a wavelength of 3 cm. A signal entering one end with a positive voltage will appear a half-wavelength later as a negative voltage. When the voltage or current is measured at each end of the wire, or along it, a different value will be found at different points along the line. Solid wire interconnects can no longer be considered internal to the node, and the current entering one end of the wire is not necessarily equal to the current leaving the other end. The wire must be considered as a distributed element and modeled as many nodes distributed along the wire. This is called a transmission line.

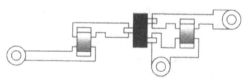

Figure 3-1 *A circuit with two components connected with a circuit trace.*

In this section, we will consider uniform transmission lines only. A uniform transmission line is a propagation medium in which wavelength, impedance and loss are constant along the entire length of the line. We will begin by modeling the transmission line as a series of circuit elements. Next, these circuit elements will be used to find the characteristic impedance, loss-per-unit length, and speed of propagation of the transmission line.

Before studying transmission lines, we will define the way we express voltage and current in a circuit. Consider a sinusoidal voltage source where the voltage is given by

$$V = V_o \cos(2\pi f t) \qquad 3.1$$

where f is the frequency of the source voltage and t is time. Equation 3.1 describes a voltage that oscillates in time from positive to negative in a sinusoidal fashion. Suppose there is a second part, an imaginary component, to this voltage that helps in the analysis of the circuit and networks. By superposition, we can add this imaginary signal to the real signal and then remove it (or ignore it) when it is no longer needed. This imaginary part is the sin component of the voltage. We can write

$$V = V_o \left(\cos(2\pi f t) + j \sin(2\pi f t) \right) = V_o e^{j 2\pi f t} = V_o e^{j\omega t} \qquad 3.2$$

As will be shown, this exponential representation is a convenient way to express the voltage and current on a transmission line. Now that we have the basic equation of the voltage in the time domain, we can combine it with the effects of transmission line length. We add an imaginary component to the exponent to describe the way voltage changes at different points along the transmission line given by x.

$$V = V_o e^{j\omega t - j\beta x} \qquad 3.3$$

One other term is needed to characterize the loss (or gain) in a signal along the length of a transmission line. The general expression for voltage on a transmission line is

$$V = V_o e^{-\gamma + j\omega t} \qquad 3.4$$

where

$$\gamma = x(j\beta + \alpha) \qquad 3.5$$

In Equation 3.5, the frequency is expressed in radians/second, or $2\pi f$, and α is the loss per wavelength. This describes a wave propagating in one direction along the transmission line. When β is positive, the wave moves along the transmission line in the positive x direction. This is called the forward-propagating wave. We can define a wave that travels backward along the transmission line by this method.

$$V = V_r e^{j\omega t + x(j\beta + \alpha)} \qquad\qquad 3.6$$

Waves can propagate in both directions simultaneously. The total voltage on the transmission line is the superposition of the forward- and backward-traveling waves given by

$$V = \left(V_f e^{-\gamma} + V_r e^{\gamma}\right)e^{j\omega t} \qquad\qquad 3.7$$

Figure 3-2(a) shows the forward-traveling wave at time t along the transmission line, and Figure 3-2(b) shows the reflected wave at time t. Notice the sinusoidal variation of the voltage with distance along the line and the attenuation of the wave as it propagates. Current along the transmission line is expressed in the same way (V_s is replaced with I_s) and is separated into forward-traveling current waves and reverse current waves.

$$I = \left(I_f e^{-\gamma} + I_r e^{\gamma}\right)e^{j\omega t} \qquad\qquad 3.8$$

The wavelength denotes the distance along the line where the voltage goes through one complete cycle, i.e., for example, from one voltage peak to the next. In terms of the voltage vector V, the wavelength is the distance Δx, which the phase of the wave vector goes 360 degrees, or 2π radians.

$$\lambda = \frac{2\pi}{\beta} \qquad\qquad 3.9$$

The propagation velocity is the distance the wave travels (Δx) along the transmission line in a certain time (Δt). If we look at the voltage at time $t + \Delta t$ along the transmission line, we must find a location $x + \Delta x$ where the voltage has passed through 2π of phase change.

$$\omega \Delta t = 2\pi = \beta \Delta x \qquad\qquad 3.10$$

The change in distance divided by the equivalent change in time is velocity.

$$v = \frac{\omega}{\beta} \qquad\qquad 3.11$$

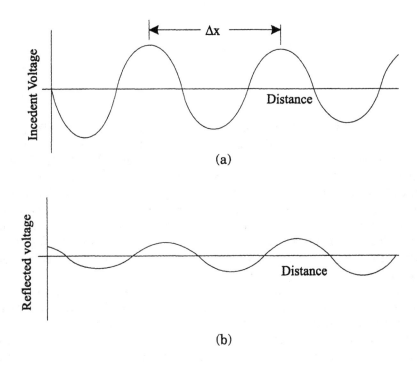

Figure 3-2 *The forward (a) and reflected (b) voltage waves along the transmission line at an instant in time.*

Example 3.1: Given a voltage of 2cos(4t + 3x) on a transmission line running in the x direction, find the wavelength and propagation velocity of the signal. If the current is $I = .5\cos(4t + 3x + \pi)$, what is the impedance of the transmission line at this frequency?

$$V = 2\cos\left(4t + 3x\right)$$

$$V = 2e^{j4t + j3x}$$

$$\omega = 4 \text{ rad/s}$$

$$\beta = 3$$

$$\lambda = \frac{2\pi}{3}$$

$$v = \frac{4}{3}$$

$$Z = \frac{2e^{j4t+j3x}}{0.5e^{j4t+j3x+j\pi}}$$

$$Z = 4e^{j\pi} = -4$$

From this point in our discussion, we will assume that there is a frequency term ($e^{j\omega t}$) associated with any voltage or current on a transmission line. Since we know there is a frequency term associated with all of these quantities, we will not write $e^{j\omega t}$ in every equation. It will be easier to work with voltage and current if we do not have to put $e^{j\omega t}$ every time. However, we must keep in mind that a frequency term accompanies all of the voltage and current terms. Some authors give these variables a special font or put a tilde over the letter. In this book, we will simply ignore the frequency term until we determine the value of a component such as a capacitor or an inductor.

A transmission line can be represented by an equivalent circuit that accounts for its distributed nature. We start with a small segment of transmission line that has a length of Δx. The equivalent circuit of a transmission line with a series resistance $R\Delta x$, an inductance $L\Delta x$, a shunt conductance of $G\Delta x$ and a capacitance of $C\Delta x$ is shown in Figure 3-3. If the length of the transmission line element Δx is small, Kirchhoff's laws will apply.

$$V(x - \Delta x) = V(x) - I(x)(sL + R)\Delta x \qquad 3.12$$

$$I(x - \Delta x) = I(x) - V(x)(sC + G)\Delta x \qquad 3.13$$

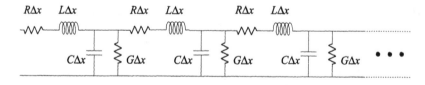

Figure 3-3 *Lumped component equivalent of a transmission line.*

If we divide both sides of the equation by Δx, rearrange the terms and take the limit as Δx approaches zero, we receive two differential equations in the form

$$-(sC + G)V(x) - (sL + R)I(x) = \frac{\partial I(x)}{\partial x} \qquad 3.14$$

$$-(sL+R)I(x) = \frac{\partial V(x)}{\partial x} \qquad \text{3.15}$$

It is no coincidence that the equations for the total voltage and current on a transmission line, Equations 3.7 and 3.8, are solutions to these differential equations. When these formulas are used, Equations 3.14 and 3.15 become

$$I = I_f e^{-\gamma} + I_r e^{\gamma} \qquad \text{3.16}$$

$$V = V_f e^{-\gamma} + V_r e^{\gamma} \qquad \text{3.17}$$

$$-(sL+R)\left(I_f e^{-\gamma} + I_r e^{\gamma}\right) = -\mathcal{W}_f e^{-\gamma} + \mathcal{W}_r e^{\gamma} \qquad \text{3.18}$$

$$-(sL+R)\left(V_f e^{-\gamma} + V_r e^{\gamma}\right) = -\mathcal{A}_f e^{-\gamma} + \mathcal{A}_r e^{\gamma} \qquad \text{3.19}$$

These equations must hold everywhere along the transmission line for all x.

$$(sL+R)I_f = \mathcal{W}_f \qquad \text{3.20}$$

$$(sC+G)V_f = \mathcal{A}_f \qquad \text{3.21}$$

Solving Equations 3.20 and 3.21 for γ yields the propagation constant in terms of the circuit elements.

$$\gamma = \sqrt{(sL+R)(sC+G)} \qquad \text{3.22}$$

The characteristic impedance of a transmission line is the ratio of the forward-traveling voltage wave and forward-traveling current wave.

$$Z_o = \frac{V_f}{I_f} \qquad \text{3.23}$$

$$= \frac{sL+R}{\gamma} \qquad \text{3.24}$$

$$= \sqrt{\frac{sL+R}{sC+G}} \qquad \text{3.25}$$

Characteristic impedance can also be found from the reverse traveling voltage and current waves.

$$\frac{V_r}{I_r} = \frac{-(sL+R)}{\gamma} = -Z_o \qquad 3.26$$

Example 3.2: A transmission line has a resistance of 2 ohms/m, an inductance of 1.5 microhenry/meter, 0.1 mhos/meter and 0.10 micofarid/meter. Find the characteristic impedance, loss, and propagation velocity at 1 MHz. The characteristic impedance is found using Equation 3.25.

$$Z_o = \sqrt{\frac{j2\pi(1E6)(1.5E-6)+2}{j2\pi(1E6)(1E-7)+0.1}}$$

$$Z_o = 3.924 - j0.7282\Omega$$

The propagation constant is found from Equation 3.22.

$$\gamma = \sqrt{\left(j2\pi(1E6)(1.5E-6)+2\right)\left(j2\pi(1E6)(1E-7)+0.1\right)}$$

$$\gamma = 0.4517 + j2.434$$

The loss is the real part of the propagation constant,

$$\alpha = 0.4517$$

and the propagation velocity is found using Equation 3.11.

$$v = \frac{2\pi(1E6)}{2.434} = 2.581 \text{ m}/\text{s}$$

Impedance is defined as the ratio of voltage and current traveling in the same direction on a transmission line. Another important quantity is the *reflection coefficient*, which is the ratio of forward- and reverse-traveling waves.

$$\Gamma = \frac{V_f}{V_r} \qquad 3.27$$

The reflected wave normally will be less than the incident, or forward, wave. The reflection coefficient will be a complex number with a magnitude of 1.0 or less. Because the forward and reflected voltage may differ in phase, the reflection coefficient will have an argument, or phase angle component.

In this section, we have postulated the existence of a forward- and reverse-traveling wave regardless of where these waves originate. There are two ways that these waves are created — the simplest being two voltage or current sources on either end of the transmission line. A more interesting and more common source of these two waves is the use of one source, or generator, and a circuit element that causes a mismatch on the transmission line. Consider the case where a transmission line is leading to a circuit element Z_L, as shown in Figure 3-4. The voltage and current at this termination, or load, is related by Ohm's law.

$$Z_L = \frac{V}{I} = \frac{V_f \, e^{-\gamma} + V_r e^{\gamma}}{I_f e^{-\gamma} - I_r e^{\gamma}} \qquad\qquad 3.28$$

$$= \frac{V_f \, e^{-\gamma} + V_r e^{\gamma}}{\dfrac{V_f}{Z_o} e^{-\gamma} - \dfrac{V_r}{Z_o} e^{\gamma}} \qquad\qquad 3.29$$

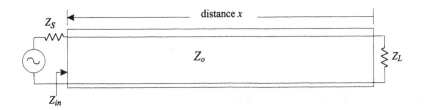

Figure 3-4 *Transmission line with a characteristic impedance of Z_o terminated with a load impedance of Z_L.*

When the location of the load impedance is at $x = 0$, the plane of reference,

$$Z_L = Z_o \frac{V_f + V_r}{V_f - V_r} \qquad\qquad 3.30$$

Solving Equation 3.30 for V_f and V_r, the reflection coefficient in terms of the characteristic impedance of the transmission line (Z_o) and the load impedance (Z_L) can be found to be

$$\Gamma = \frac{Z_L - Z_o}{Z_L + Z_o} \qquad\qquad 3.31$$

At other points along the transmission line the voltage and current of the waves are different. The reflection coefficient changes as we move along the transmission line. This is demonstrated in Equation 3.29, which can be rewritten in terms of Z_o and Z_L

$$Z_{in} = Z_o \frac{Z_L \cosh \gamma \ x + Z_o \sinh \gamma \ x}{Z_o \cosh \gamma \ x + Z_L \sinh \gamma \ x} \qquad 3.32$$

where Z_{in} is the input impedance and x is the length of the transmission line. If the transmission line is lossless, Equation 3.32 becomes

$$Z_{in} = Z_o \frac{Z_L + j Z_o \tan \beta \ x}{Z_o + j Z_L \tan \beta \ x} \qquad 3.33$$

The reflection coefficient along the transmission line changes as we move along the line. Using the forward- and reverse-traveling waves from Equation 3.17 in Equation 3.27, the reflection coefficient at location x is

$$\Gamma(x) = \Gamma_o e^{-2\gamma x} \qquad 3.34$$

where Γ_o is the reflection coefficient at $x = 0$. When the transmission line is lossless,

$$\Gamma(x) = \Gamma_o (\cos 2\beta x - j \sin 2\beta x) \qquad 3.35$$

When moving along a lossless transmission line, the magnitude of the reflection coefficient stays constant. The argument, or angle, of $\Gamma(x)$ varies as exp($-2\beta x$) as the reference plane moves in the x direction.

The reflection coefficient is frequently used in the analysis and design of microwave circuits to cancel unwanted reflections or cause a specific circuit response. These circuits are designed using components or circuit elements that cause reflections on a transmission line. The reflection coefficient is such a valuable tool that a method of graphing impedances and admittances in relation to the reflection coefficients they create has been developed.

Example 3.3: Figure 3-5 shows a transmission line with a characteristic impedance of 75 ohms that is terminated by a 100 ohm load. Find the reflection coefficient at any point along the transmission line. The reflection coefficient at the end of the transmission line where $x = 0$ is

$$\Gamma_o = \frac{100 - 75}{100 + 75} = \frac{25}{175} = 0.143$$

The reflection coefficient at any value of x is

$$\Gamma(x) = 0.143\,(\cos 2\beta x - j\sin 2\beta x)$$

In terms of wavelengths we can write

$$\Gamma(x) = 0.143\left(\cos\frac{4\pi}{\lambda}x - j\sin\frac{4\pi}{\lambda}x\right)$$

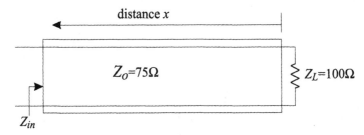

Figure 3-5 *The transmission line for Example 3.3.*

3.3 Introduction to the Smith Chart

The Smith chart is a graphical tool used to visualize, analyze, and design high-frequency circuits. It was invented by Phillip Smith [5] and has become an indispensable tool for microwave engineers. Figure 3-6 shows a Smith chart that has constant impedance lines. Two sets of circles can be seen on the Smith chart. In one set, the circles are complete and nested, one inside the other, all touching at the far right-hand side of the chart. These are circles of constant resistance, which will be discussed later in this section. The other set consists of partial circles, all touching the horizontal centerline of the chart. These are constant reactance circles, or plots, of the constant imaginary part of impedance.

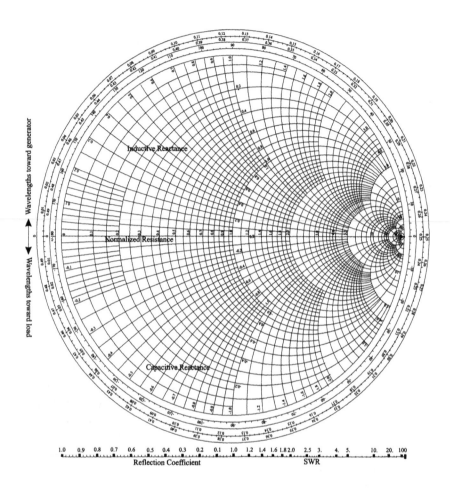

Figure 3-6 *The Smith chart plotting constant impedance circles.*

Figure 3-7 shows a Smith chart with constant admittance lines. On the admittance Smith chart, the full circles are the constant conductance lines. The partial circles are the constant susceptance lines. One common characteristic of the admittance and the impedance Smith chart is the location of the open and short on both charts. In fact, a reflection coefficient will be plotted in the same place on both charts.

If a rectangular grid is superimposed onto the Smith chart, the real and imaginary part of the reflection coefficient can be plotted in Cartesian coordinates. The horizontal axis is the real part of the reflection coefficient, and the vertical axis is the imaginary part. The reflection coefficient equals zero at the center of the Smith chart. The edge of the Smith chart corresponds to all points where the reflection coefficient is equal to 1. When a reflection coefficient is expressed as a magnitude A and angle α, it is plotted on the Smith chart at a radius of A from the center of the chart, at an angle α from the positive x-axis.

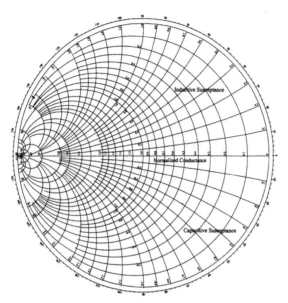

Figure 3-7 *The Smith chart plotting constant admittance circles.*

The simplest and most common way to use a Smith chart is to convert a load impedance to a reflection coefficient and vice versa. The reflection coefficient is a function of characteristic impedance Z_o and the load impedance. When we normalize the load impedance by the characteristic impedance, the reflection coefficient becomes

$$\Gamma = \frac{Z-1}{Z+1}$$

3.36

where $Z = Z_L/Z_o$. Constant resistance circles appear on the Smith chart as circles that are always centered on the real axis and touch the same point at the far right side of the chart. With $\Gamma = u + jv$ and $Z = x + jy$, Equation 3.36 becomes

$$u + jv = \frac{x + jy - 1}{x + jy + 1}$$

3.37

which can be written as

$$jy + x - jv + vy - jvx - u - juy - ux = 1 \qquad 3.38$$

Equation 3.34 is separated into the real and imaginary parts:

$$jy - jv - jvx - juy = 0 \qquad 3.39$$

and

$$x + vy - u - ux = 1 \qquad 3.40$$

We first solve Equations 3.39 and 3.40 for y.

$$y = \frac{v(1+x)}{1-u} \qquad 3.41$$

We then solve for x by substituting Equation 3.41 back into Equations 3.39 and 3.40 so that it too can be eliminated.

$$v^2 = \left(u - \frac{x}{1-x}\right)^2 = \frac{x^2 = x-1}{(x=1)^2}$$

$$\qquad 3.42$$

$$(u-1)^2 = \left(v - \frac{1}{2y}\right)^2 = \frac{1}{4}y^2$$

This exercise yields the equation of a circle.

$$(x-a)^2 + (y-b)^2 = r^2$$

where the center of the circle is at $x = a$, $y = b$, and r is the radius. By comparing the equation of a circle to Equation 3.42, circles of constant resistance are centered at $b = 0$ in all cases, and the radius is always equal to a. In other words, the circles are always centered on the x-axis. Since the radius and the x location of the center are the same, all the resistance circles will meet on the x-axis when $\Gamma = 1$. Another equation of a circle appears that is centered at $a = 1$ in all cases and that r is equal to b.

   Example 3.4: Using the Smith chart in Figure 3-6 with a characteristic impedance of 50 ohms, plot the following load impedances:
*   A. $120 - j75$ ohms,
*   B. $30 + j120$,
*   C. a 50 ohm load with any possible imaginary part,
*   D. a $-j30$ ohm load with any possible real part.

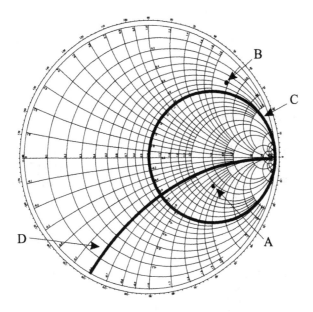

Figure 3-8 *The impedances in Example 3.4 are plotted on the 50 ohm Smith chart.*

The Smith chart is valuable for plotting impedance along a transmission line. Consider the transmission line shown in Figure 3-5. The reflection coefficient at the load (when $x = 0$) is not matched to the transmission line impedance. We can easily plot this point on the Smith chart. Let us begin to add length to the transmission line so that the distance increases. If the transmission line is lossless, Equation 3.35 shows that the magnitude of the reflection coefficient remains constant and only the angle changes. When $\Gamma(x)$ is plotted on the Smith chart, the point moves in a circle, centered in the middle of the Smith chart, and rotates in the clock-wise direction. The arrow in Figure 3-6 labeled "toward the generator" shows the direction of rotation as transmission line length is added. If the transmission line is shortened, the reflection coefficient, as plotted on the Smith chart, rotates counter-clockwise. This direction is shown in Figure 3-6 by the arrow labeled "toward the load."

Example 3.5: Figure 3-9 shows two transmission lines with one of the lines terminated. Plot Z_{in} on the Smith chart. A load with an impedance of $45 - j20$ ohms is connected to a 50 ohm transmission line that is one-eighth of a wavelength long. Then, the 50 ohm transmission line is connected to a 75 ohm transmission line that is three-eights of a wavelength long. Find the input impedance of the circuit using the Smith chart.

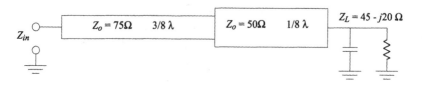

Figure 3-9 *Circuit for Example 3.5.*

Since the transmission line on the right is a 50 ohm line, we begin with a 50 ohm Smith chart. The termination impedance of $45 - j20$ is plotted at a point labeled Z_L on the Smith chart in Figure 3-10(a). The impedance is transformed along path B by the 45 degrees long transmission line ending at point C.

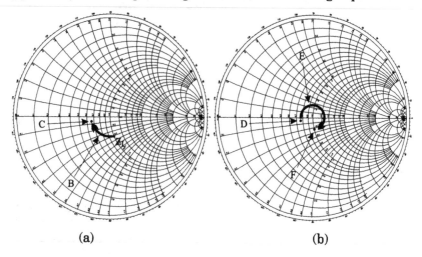

Figure 3-10 *The 50 ohm Smith chart (a) and the 75 ohm Smith chart (b) for Example 3.5.*

The next transmission line is a 75 ohm characteristic impedance; therefore, point C is transferred to a 75 ohm Smith chart in Figure 3-10(b). The impedance at point C is found by multiplying the reading on the Smith chart in Figure 3-10(a) by 50 ohms. The impedance is then divided by 75 ohms and plotted on the 75 ohm Smith chart in Figure 3-10(b) at point D. Finally, the reflection coefficient at point D is transformed around the Smith chart by the last transmission line, shown as path E, which is 3/8 of a wavelength long. The input impedance and reflection coefficient of the circuit is plotted at point F.

3.4 Wave Vectors and S-Parameters

At high frequencies it is difficult to measure the voltage and current at a terminal of a device or network. On the other hand, directional couplers can easily measure the power flow into or out of a circuit. Hence, it is suitable to describe the electrical properties of a circuit by some means of power flow or ratios thereof.

Concerning the incident voltage and current in a transmission line, it would be more convenient to normalize these by the characteristic impedance (Z_o). We can define a normalized incident vector a in terms of the incident voltage and current.

$$a = \frac{V_f}{\sqrt{Z_o}} = I_f \sqrt{Z_o} = \frac{V_f + Z_o I_f}{2\sqrt{Z_o}} \qquad 3.43$$

We can also define a normalized reflected vector b, which is a function of the reflected voltage and current.

$$b = \frac{V_r}{\sqrt{Z_o}} = -I_r \sqrt{Z_o} = \frac{V_r - Z_o I_r}{2\sqrt{Z_o}} \qquad 3.44$$

In terms of a and b, the incident power is

$$P^+ = |a|^2 \qquad 3.45$$

The reflected power is

$$P^- = |b|^2 \qquad 3.46$$

The reflection coefficient is

$$\Gamma = \frac{b}{a} \qquad 3.47$$

The actual power delivered to the load will be

$$P = |a|^2 - |b|^2 \qquad 3.48$$

and the VSWR of the termination is

$$\rho = \frac{|a| + |b|}{|a| - |b|} \qquad 3.49$$

With the new set of variables (a and b) it is simpler to compare the calculated circuit response to the power flow measured with directional couplers and power meters.

Usually, there is very little interest in the actual values of a and b. Rather, the interest lies in their ratio, which defines reflection and transmission of power in a network. The reflection coefficient Γ in Equation 3.47 is the S-parameter representation of a one port. Just as with Z- and Y-parameters, S-parameters can be applied to n-port networks by defining incoming and outgoing waves at each port:

$$a_i = \frac{1}{2}\left(\frac{V_i + Z_o I_i}{\sqrt{Z_o}} \right)$$

3.50

$$b_j = \frac{1}{2}\left(\frac{V_j - Z_o I_j}{\sqrt{Z_o}} \right)$$

3.51

where i and j are the port number. For the two-port network shown in Figure 3-7, the S-parameters are defined as

$$S_{11} = \frac{b_1}{a_1}\bigg|_{a_2=0}$$

3.52

$$S_{12} = \frac{b_1}{a_2}\bigg|_{a_1=0}$$

3.53

$$S_{21} = \frac{b_2}{a_1}\bigg|_{a_2=0}$$

3.54

$$S_{22} = \frac{b_2}{a_2}\bigg|_{a_1=0}$$

3.55

The power wave leaving Port 1 is the sum of the power reflected from the input port ($S_{11}a_1$) plus the power passing through the two-port from Port 2 to Port 1 ($S_{21}a_2$).

$$b_1 = S_{11}a_1 + S_{12}a_2$$

3.56

Correspondingly, the power wave leaving Port 2 is

$$b_2 = S_{21}a_1 + S_{22}a_2 \qquad 3.57$$

or, in matrix representation,

$$\begin{bmatrix} a_1 \\ a_2 \end{bmatrix} = \begin{bmatrix} S_{11} & S_{12} \\ S_{21} & S_{22} \end{bmatrix} \begin{bmatrix} b_1 \\ b_2 \end{bmatrix} \qquad 3.58$$

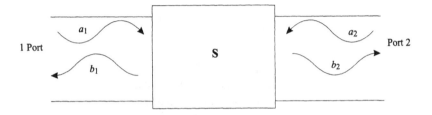

Figure 3-11 *A two-port network and the incident vectors a_1 and a_2 and outgoing waves b_1 and b_2 at the two ports.*

Since the waves a_i and b_i are functions of voltage and current vectors, the S-parameters can be related to the Z-, Y-, and ABCD-parameters of the network. Let us look at the matrix of S-parameters and column vectors **a'** and **b'** in Equation 3.58

$$\mathbf{b'} = \mathbf{Sa'} \qquad 3.59$$

where

$$\mathbf{a'} = \frac{1}{\sqrt{Z_o}}(\mathbf{V} + Z_o\mathbf{I}) \qquad 3.60$$

$$\mathbf{b'} = \frac{1}{\sqrt{Z_o}}(\mathbf{V} - Z_o\mathbf{I}) \qquad 3.61$$

and **V** and **I** are the column vectors consisting of the voltages on the two-port terminals and the current flowing into the terminals. The impedance matrix of the two port is defined as

$$\mathbf{V} = \mathbf{ZI} \qquad 3.62$$

Inserting Equation 3.62 into Equations 3.60 and 3.61 yields

$$\mathbf{a'} = \frac{1}{\sqrt{Z_o}}\left(\mathbf{ZI} + Z_o\mathbf{I}\right) \qquad 3.63$$

$$\mathbf{b'} = \frac{1}{\sqrt{Z_o}}\left(\mathbf{ZI} - Z_o\mathbf{I}\right) \qquad 3.64$$

We find that the S-parameter matrix can be expressed as a function of the Z-parameter matrix.

$$\mathbf{S} = \frac{\mathbf{a'}}{\mathbf{b'}} = \frac{\mathbf{ZI} - Z_o\mathbf{I}}{\mathbf{ZI} + Z_o\mathbf{I}} = \left(\mathbf{Z} - Z_o\right)\left(\mathbf{Z} + Z_o\right)^{-1} \qquad 3.65$$

Note that the constant Z_o in Equation 3.60 was converted to a diagonal matrix by multiplying Z_o by the identity matrix. S-parameters can be related to other network parameters by performing similar derivations to the one above.

Example 3.6: Consider a 75 ohm transmission line that is terminated with a $35 + j120$ ohm load. We want to find the reflection coefficient at the load and the VSWR using the wave vectors.

Normalizing the voltage source to 1, the incident voltage and current are (as if the transmission line were perfectly matched)

$$V^+ = \frac{1}{2} \qquad\qquad I^+ = \frac{1}{150}$$

The reflected voltage and current are found to be

$$V^- = \frac{75 - (35 + j120)}{2(75 + (35 + j120))} = \frac{40 - j120}{220 + j240} = \frac{4}{-5 + j9}$$

$$I^- = \frac{75 - (35 + j120)}{150(75 + (35 + j120))} = \frac{4}{-375 + j675}$$

The vectors a and b are

$$a = \frac{1}{2\sqrt{75}} \qquad\qquad b = \frac{4}{\sqrt{75}(-5 + j9)}$$

and the reflection coefficient and the VSWR are

$$\Gamma = S_L = \frac{8}{-5+j9}$$

$$\rho = \frac{\dfrac{1}{2}+\dfrac{4}{\sqrt{106}}}{\dfrac{1}{2}-\dfrac{4}{\sqrt{106}}} = 7.97:1$$

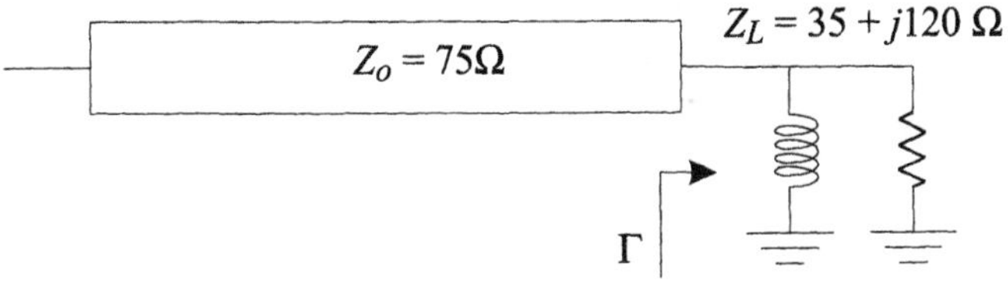

Figure 3-12 *The circuit for Example 3.6 with $Z_o = 75$ ohms and $Z_L = 35 + j120$ ohms.*

Example 3.7: Consider the circuit shown in Figure 3-13. The termination in Example 3.6 is connected to the same 75 ohm source through a transmission line 3/8 λ long. At the load, the parameters calculated above are the same. At the generator, the length of the transmission line has shifted the phases. In terms of incident voltage and current at the generator,

$$V_g^+ = \frac{1}{2}\exp\left(j2\pi\frac{3}{8}\right) = -\frac{1}{2\sqrt{2}}+j\frac{1}{2\sqrt{2}}$$

$$I_g^+ = \frac{1}{150}\exp\left(j2\pi\frac{3}{8}\right) = \frac{1}{150}\left(-\frac{1}{\sqrt{2}}+j\frac{1}{\sqrt{2}}\right)$$

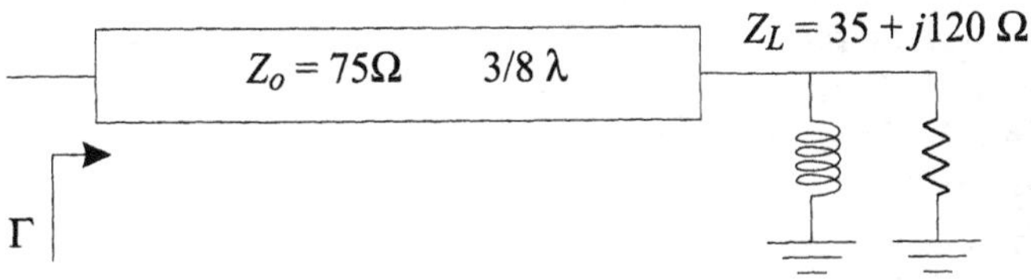

Figure 3-13 *The circuit of Example 3.7 with a transmission line 3/8 of a wavelength long.*

The reflected voltage and current are

$$V_g^- = \frac{4}{-5+j9}\exp\left(-j2\pi\frac{3}{8}\right) = \frac{2\sqrt{2}+j2\sqrt{2}}{5-j9}$$

$$I_g^- = \frac{4}{-375 + j675} \exp\left(-j2\pi\frac{3}{8}\right) = \frac{2\sqrt{2} + j2\sqrt{2}}{375 - j675}$$

The vectors a and b are

$$a' = \frac{1}{2\sqrt{75}}\left(-\frac{1}{\sqrt{2}} + j\frac{1}{\sqrt{2}}\right)$$

$$b' = \frac{2\sqrt{2} + j2\sqrt{2}}{\sqrt{75}(+5 - j9)}$$

The reflection coefficient and the VSWR are

$$S_{in} = \Gamma_{in} = \frac{b'}{a'} = \frac{\dfrac{4}{\sqrt{75}(-5 + j9)}\exp\left(-j2\pi\dfrac{3}{8}\right)}{\dfrac{1}{2\sqrt{75}}\exp(j2\pi\dfrac{3}{8})} = \frac{8}{-5 + j9}\exp\left(-j4\pi\frac{3}{8}\right)$$

$$= \frac{8\exp(-j\pi\dfrac{3}{2})}{-5 + j9} = \frac{j8}{5 - j9}$$

$$\rho = 7.97:1 \qquad \left(\left|-\frac{1}{\sqrt{2}} + j\frac{1}{\sqrt{2}}\right| = 1\right)$$

Note that the phase of the reflection coefficient has been shifted by two times the electrical length because the waves must reach the load and then return through the length of transmission line.

Example 3.8: Consider a two-port network with Port 2 terminated, as shown in Figure 3-14. Suppose we want to know the input reflection coefficient Γ_i. The ratio of the reflected wave from the load (a_2) to the incident wave on the load (b_2) is

$$\Gamma_L = \frac{a_2}{b_2}$$

The incident wave on the output of the two-port is

$$a_2 = \Gamma_L b_2$$

Substituting this in Equation 3.56 and 3.57, the response of the two-port is given by

$$b_1 = S_{11}a_1 + S_{12}\Gamma_L b_2$$

$$b_2 = S_{21}a_1 + S_{22}\Gamma_L b_2$$

By solving the second equation for b_2

$$b_2 = \frac{S_{21}a_1}{1 - S_{22}\Gamma_L}$$

we can solve the first equation and find the input reflection coefficient.

$$\Gamma_i = \frac{b_1}{a_1} = S_{11} + \frac{S_{12}S_{21}\Gamma_L}{1 - S_{22}\Gamma_L}$$

The S-parameters can be manipulated algebraically just as any of the other linear parameters.

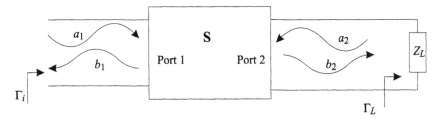

Figure 3-14 *The terminated two-port in Example 3.8.*

This has been a discussion of the basics of S-parameters and their relation to the other linear network parameters. The voltage across a load or network port is split into two components. One component is the voltage that would exist if the load or port were perfectly matched to the transmission line (otherwise known as the incident voltage), and the second component is the reflected voltage, or difference between the actual and incident voltages. The concept is the same for current. Voltage and current are the defining quantities of the incident and reflected waves a and b.

This discussion has simplified a few concepts. Most applications have a characteristic impedance of 50 ohms. In general, however, the characteristic impedance on the ith port of a network can be any complex number Z_{si} other than 0 or infinity. In this case, we must define the column vectors $\boldsymbol{a}'$ and $\boldsymbol{b}'$ for an n-port as

$$\boldsymbol{a}' = \boldsymbol{R_s}\left(\boldsymbol{V} + \boldsymbol{Z_s}\boldsymbol{I}\right) \qquad \text{3.66}$$

$$\boldsymbol{b}' = \boldsymbol{R_s}\left(\boldsymbol{V} + \boldsymbol{Z_s}^*\boldsymbol{I}\right) \qquad \text{3.67}$$

where R_s and Z_s are diagonal matrices whose i^{th} components are given by

$$\frac{1}{2}\frac{1}{\sqrt{\text{Re}(Z_{si})}} \qquad \text{3.68}$$

and Z_{si} respectively. We now derive new formulas for V, I and the conversion formulas for Y- and Z-parameters. Using the basics given in this chapter, we can now discuss power gain, insertion loss, load matching, and stability, categories which are needed for circuit design.

The following list of equations allows us to convert between two-port S-parameters and Z-, Y-, and chain parameters:

$$S_{11} = \frac{(z_{11} - 1)(z_{22} + 1) - z_{12}z_{21}}{(z_{11} + 1)(z_{22} + 1) - z_{12}z_{21}} \qquad \text{3.69}$$

$$= \frac{(1 - y_{11})(1 + y_{22}) + y_{12}y_{21}}{(1 + y_{11})(1 + y_{22}) - y_{12}y_{21}} \qquad \text{3.70}$$

$$= \frac{Z_o A + B - C(Z_o)^2 - Z_o D}{Z_o A + B + C(Z_o)^2 + Z_o D} \qquad \text{3.71}$$

$$S_{12} = \frac{2z_{12}}{(z_{11} + 1)(z_{22} + 1) - z_{12}z_{21}} \qquad \text{3.72}$$

$$= \frac{-2y_{12}}{(1 + y_{11})(1 + y_{22}) - y_{12}y_{21}} \qquad \text{3.73}$$

$$= \frac{2Z_o(AD - BC)}{Z_o A + B + C(Z_o)^2 + Z_o D} \qquad \text{3.74}$$

$$S_{21} = \frac{2z_{21}}{(z_{11} + 1)(z_{22} + 1) - z_{12}z_{21}} \qquad \text{3.75}$$

$$= \frac{-2y_{21}}{(1+y_{11})(1+y_{22}) - y_{12}y_{21}} \qquad 3.76$$

$$= \frac{2Z_o}{Z_oA + B + C(Z_o)^2 + Z_oD} \qquad 3.77$$

$$S_{22} = \frac{(z_{11}+1)(z_{22}-1) - z_{12}z_{21}}{(z_{11}+1)(z_{22}+1) - z_{12}z_{21}} \qquad 3.78$$

$$= \frac{(1+y_{11})(1-y_{22}) + y_{12}y_{21}}{(1+y_{11})(1+y_{22}) - y_{12}y_{21}} \qquad 3.79$$

$$= \frac{-Z_oA + B - C(Z_o)^2 + Z_oD}{Z_oA + B + C(Z_o)^2 + Z_oD} \qquad 3.80$$

In Equations 3.69 to 3.79, $z_{ij} = Z_{ij}/Z_o$ and $y_{ij} = Z_o Y_{ij}$.

$$z_{11} = \frac{(1+S_{11})(1-S_{22}) + S_{12}S_{21}}{(1-S_{11})(1-S_{22}) - S_{12}S_{21}} \qquad 3.81$$

$$z_{12} = \frac{2S_{12}}{(1-S_{11})(1-S_{22}) - S_{12}S_{21}} \qquad 3.82$$

$$z_{21} = \frac{2S_{21}}{(1-S_{11})(1-S_{22}) - S_{12}S_{21}} \qquad 3.83$$

$$z_{22} = \frac{(1-S_{11})(1+S_{22}) + S_{12}S_{21}}{(1-S_{11})(1-S_{22}) - S_{12}S_{21}} \qquad 3.84$$

$$y_{11} = \frac{(1-S_{11})(1+S_{22}) + S_{12}S_{21}}{(1+S_{11})(1+S_{22}) - S_{12}S_{21}} \qquad 3.85$$

$$y_{12} = \frac{-2S_{12}}{(1+S_{11})(1+S_{22}) - S_{12}S_{21}} \qquad 3.86$$

$$y_{21} = \frac{-2S_{21}}{(1+S_{11})(1+S_{22}) - S_{12}S_{21}} \qquad 3.87$$

$$y_{22} = \frac{(1+S_{11})(1-S_{22})+S_{12}S_{21}}{(1+S_{11})(1+S_{22})-S_{12}S_{21}}$$ 3.88

$$A = \frac{(1+S_{11})(1-S_{22})+S_{12}S_{21}}{2S_{21}}$$ 3.89

$$B = Z_o \frac{(1+S_{11})(1+S_{22})-S_{12}S_{21}}{2S_{21}}$$ 3.90

$$C = \frac{(1-S_{11})(1-S_{22})-S_{12}S_{21}}{Z_o 2S_{21}}$$ 3.91

$$D = \frac{(1-S_{11})(1+S_{22})+S_{12}S_{21}}{2S_{21}}$$ 3.92

S-parameters and transmission lines are the two main tools of the microwave engineer. Manufacturers measure transistors and publish the S-parameters in data sheets. Transistors are usually built into a circuit with transmission lines and other components. All these parts are designed to work in unison to amplify a signal by a predictable amount. Many other types of circuits are designed to filter, radiate, modulate, demodulate, or control microwave signals. There are a number of different types of transmission lines that are useful in certain applications.

3.5 Microstrip Transmission Lines

There are many physical transmission line structures used in microwave circuits; common transmission lines include waveguide, fiber optic cable, and coaxial cable. This section covers in some detail the most common type of transmission lines used in microwave amplifiers: the microstrip line.

Figure 3-15 shows some of the transmission structures used in microwave circuits. Waveguides are common in high-power and low-loss applications. Coaxial cables are frequently used to transmit signals from an antenna to an amplifier. Striplines are conducting strips sandwiched between two conducting ground planes. These strips are insulated from the ground planes by a dielectric. Voltage and current waves propagate along the strips guided by two ground planes. Characteristic impedance is mainly a function of the strip width, the dielectric constant of the substrate, and the thickness of the dielectric. Slot lines have a nonconducting slot where the wave propagates guided by two grounds on either side of the gap, as shown in Figure 3-15(d). Coplanar waveguides are similar to slotlines, but contain a conduction strip in the middle of the slot. A dielectric image line, shown in Figure 3-15(f), is common at frequencies above 60 GHz. An image line is a slab of dielectric material bonded to a conducting ground plane. The wave is guided by the surface of the slab and the ground plane.

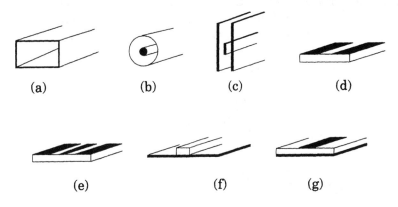

(a)   (b)   (c)   (d)

(e)   (f)   (g)

Figure 3-15 *Diagram of some microwave transmission structure: wave-guide (a), coaxial cable (b), stripline (c), slot line (d), coplanar waveguide (e), image line (f) and microstrip line (g).*

Microstrip lines are the most common form of microwave transmission lines used in amplifiers. A microstrip line is a narrow, conducting strip attached to the top of a dielectric sheet. The bottom of the sheet is covered with a conductor, serving as a ground plane. Signals propagate along the strip. The impedance of a microstrip line primarily depends on the width and thickness of the microstrip line and the dielectric constant of the material. The velocity of propagation is also related to these parameters. The width of a microstrip line is sometimes expressed as a ratio of strip width to substrate thickness, or w/h ratio. The characteristic impedance of a coaxial line can be calculated by solving the electromagnetic boundary value equations. This calculation does not work for microstrip lines because there is no closed-form solution. However, some very good approximations have been developed [1, 3]. One of these approximations is given below for calculating the characteristic impedance given the relative dielectric constant of the substrate and the w/h ratio of the microstrip line. When the microstrip is narrow ($w/h < 3.3$), the characteristic impedance is

$$Z_o = \frac{119.9}{\sqrt{2(\varepsilon_r + 1)}} \left\{ \ln\left[4\frac{h}{w} + \sqrt{16\left(\frac{h}{w}\right)^2 + 2} \right] - \frac{1}{2}\left(\frac{\varepsilon_r - 1}{\varepsilon_r + 1}\right)\left(\ln\frac{\pi}{2} + \frac{1}{\varepsilon_r}\ln\frac{4}{\pi}\right) \right\} \qquad 3.93$$

when the microstrip is wide ($w/h > 3.3$), the characteristic impedance is

$$Z_0 = \frac{119.9\pi}{2\sqrt{\varepsilon_r}} \left\{ \frac{w}{2h} + \frac{\ln 4}{\pi} + \frac{\ln(e\pi^2/16)}{2\pi}\left(\frac{\varepsilon_r-1}{\varepsilon_r^2}\right) + \frac{\varepsilon_r+1}{2\pi\varepsilon_r}\left[\ln\frac{e\pi}{2} + \ln\left(\frac{w}{2h}+0.94\right)\right] \right\}^{-1} \qquad 3.94$$

Figure 3-16 shows the characteristic impedance of a microstrip line as a function of width-to-height ratio. Each line represents the relation between impedance and w/h for different substrate dielectric constants.

It is more useful to obtain an expression for finding the w/h ratio given the desired characteristic impedance of the microstrip line and the dielectric constant. For narrow strips, $Z_0 > 44 - 2\,\varepsilon_r$

$$\frac{w}{h} = \left(\frac{\exp H'}{8} - \frac{1}{4\,\exp H'}\right)^{-1} \qquad 3.95$$

where

$$H' = \frac{Z_0\sqrt{2(\varepsilon_r+1)}}{119.9} - \frac{1}{2}\left(\frac{\varepsilon_r-1}{\varepsilon_r+1}\right)\left(\ln\frac{2}{\pi} + \frac{1}{\varepsilon_r}\ln\frac{4}{\pi}\right) \qquad 3.96$$

When the strip is wide, $Z_0 < 44 - 2\,\varepsilon_r$,

$$\frac{w}{h} = \frac{2}{\pi}\{(d-1) - \ln(2d-1)\} + \frac{\varepsilon_r-1}{\pi\varepsilon_r}\left\{\ln(d-1) + 0.293 - \frac{0.517}{\varepsilon_r}\right\} \qquad 3.97$$

where

$$d = \frac{59.95\,\pi^2}{Z_0\sqrt{\varepsilon_r}} \qquad 3.98$$

The relative dielectric constant of the microstrip line is approximated by

$$\varepsilon_{eff} = \frac{\varepsilon_r}{0.96 + \varepsilon_r(0.109 - 0.004\varepsilon_r)\{\log(10+Z_0)-1\}} \qquad 3.99$$

The propagation velocity of the signal in a microstrip line is given by

$$v_g = \frac{c}{\sqrt{\varepsilon_0\varepsilon_{eff}}} \qquad 3.100$$

Figure 3-17 shows the ratio of wavelength on the microstrip line to free space wavelength as a function of w/h and substrate dielectric.

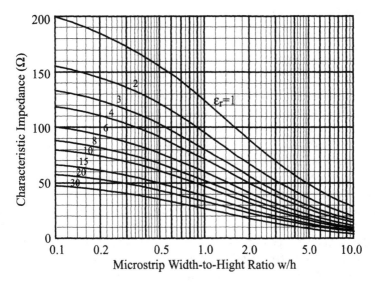

Figure 3-16 *Plot of characteristic impedance of microstrip lines.*

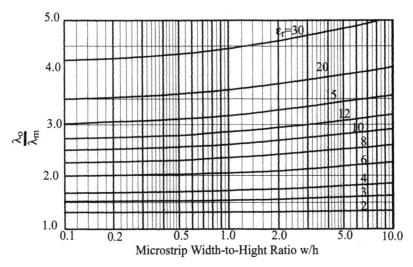

Figure 3-17 *Plot of wavelength on microstrip lines.*

One disadvantage of microstrip lines lies in the fact that they exhibit frequency-dispersive effects, i.e., the propagation characteristics of the microstrip line changes with frequency. However, this effect is negligible unless the frequency is high enough for the microstrip line or the substrate height to become an appreciable fraction of a wavelength. The relative permittivity experiences the most change as frequency increases. A few approximations have been published [2, 3, 4]; one approximation is by Edwards where h is the substrate height in millimeters.

$$\varepsilon_{eff}(f) = \varepsilon_r - \frac{\varepsilon_r - \varepsilon_{eff}}{1 + (h/Z_0)^{1.33}(0.43f^2 - 0.009f^3)} \qquad 3.101$$

When the frequency is too high, another substrate should be used for the circuit. The upper frequency limit is thought to occur when the widest line becomes approximately one-eighth of a wavelength wide. The characteristic impedance is fairly constant with frequencies up to this point.

Many microstrip circuit elements, including open-circuited transmission lines, short-circuited transmission lines, and transmission lines of differing impedance are used as tuning elements when designing microwave amplifiers. These elements are joined and bent causing discontinuities in the orderly flow of power along the transmission line. Sometimes the effects of the discontinuities must be considered. Most microwave computer aided design systems have junction elements such as a step in the width of a microstrip line, T-junctions, X-junctions, and bends, as shown in Figure 3-18.

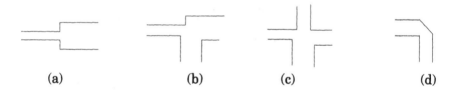

(a)　　　　　　(b)　　　　　　(c)　　　　　　(d)

Figure 3-18 *Common discontinuities in microstrip lines: a step in width (a), a junction of three microstrip lines (b), a junction of four lines (c), and a 90 degree bend with a chamfered corner (d).*

Example 3.9: Find the width of a 50 ohm line on a sapphire substrate, $\varepsilon_r = 4.2$, that is 0.65 mm thick. Since 50 is greater than 44-2 (4.2), Equations 3.95 and 3.96 are used to find the width of the 50 ohm line. First, we determine H'.

$$H' = \frac{50\sqrt{2(4.2+1)}}{119.9} - \frac{1}{2}\left(\frac{4.2-1}{4.2+1}\right)\left(\ln\frac{2}{\pi} + \frac{1}{4.2}\ln\frac{4}{\pi}\right) = 1.3905$$

We then calculate the width/height ratio.

$$\frac{w}{h} = \left(\frac{\exp 1.3905}{8} - \frac{1}{4 \exp 1.3905}\right)^{-1} = 2.2979$$

3.6   Summary

This chapter presented four valuable concepts for designing and building microwave amplifiers. First, we introduced transmission lines. Wires that interconnect components can no longer be treated as a part of the circuit node at high frequencies. Wires or circuit traces can be modeled as transmission lines with a characteristic impedance, propagation velocity, and length. In Section 3.2, we formulated the mathematical expressions for transmission lines. We found that, since the circuit interconnects must be treated as transmission lines, transmission lines can be used as the circuit elements. Also, instead of using capacitors and inductors, we can use small sections of transmission line. For example, at microwave frequencies, a short open-circuited transmission line connected in shunt can function much the same way a shunt capacitor does. In fact, if space permits, it is a preferred practice to use transmission lines wherever possible in circuit design.

Section 3.3 introduced the Smith chart, a graphical tool for analyzing and designing circuits using transmission line circuits. The Smith chart is a collection of constant impedance or admittance lines plotted on the reflection coefficient plane upon which transmission lines transform a reflection coefficient around in great circles. The Smith chart makes it easy to relate impedances to reflection coefficients and manipulate the reflection coefficients by adding transmission lines.

Further, since we will be designing microwave circuits with transmission lines, S-parameters are a convenient way to describe these circuits mathematically. Section 3.4 described the formulation of the S-parameter and how it relates to voltages and currents on a transmission line. Whereas Z-, Y-, and chain parameters are ratios of voltages and currents at the input and output of a circuit, S-parameters are ratios of traveling waves entering and leaving a circuit. These waves are fed into and extracted from a microwave circuit by transmission lines.

Finally, many different kinds of transmission lines can be used on the circuit board of a microwave amplifier. Microstrip lines are the most common type because of their ease of manufacture and low loss. Some examples of amplifiers using microstrip lines will be shown in the following chapters.

3.7   Problems

3.1   Find the wavelength, propagation velocity, and loss of a transmission line where $V = .34\ e^{j3e7t}{-}_j(35+.3)$.

3.2   Find the voltage and current along a 300 ohm transmission line with a propagation velocity of 1E8 m/s and a loss of 0.2 dB/meter.

3.3   Find the wavelength and loss of a transmission line in terms of L, C, R, and G in Equation 3.22.

3.4   Find the equivalent L, C, R and G for a 75 ohm transmission line that has a propagation velocity of 1.2E8 m/s and a loss of 0.5 dB/m.

3.5   Find the VSWR and reflection coefficient of a 75 ohm load on a 50 ohm transmission line.

3.6   What is the reflection coefficient of a 50 ohm resistor in parallel with a 100 pF capacitor terminating a 50 ohm transmission line as a function of frequency?

3.7   Solve the input impedance and reflection coefficient of a lossless 75 ohm transmission line terminated by a 300 ohm resistor as a function of transmission line length.

3.8   Plot the following impedances on a 40 ohm Smith chart:
a) $10 - j70$
b) $80 + j10$
c) $40 + j40$
d) $20 - j80$
e) $30 + j0$.

3.9. One 75 ohm transmission line is split into two 100 ohm transmission lines, one a quarter of a wavelength long and the other three-eights of a wavelength long. Both are terminated by 100 ohm loads, as shown in Figure 3-19. Find the reflection coefficient at the junction of the three transmission lines.

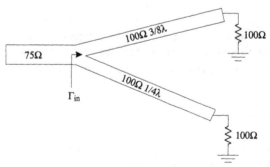

Figure 3-19 *The circuit in Problem 3.9.*

3.10 Plot the constant 30 ohm resistance line on the 50 ohm Smith chart.

3.11 Find the two-port S-parameters of a series impedance.

3.12 Find the two-port S-parameters of a shunt impedance.

3.13 Find the input reflection coefficient in a 50 ohm characteristic imped- ance network of a 40 ohm transmission line that is 1/4 wavelength long terminated with a 60 ohm resistor.

3.14 Using the low frequency approximation, find the width-to-height ratio of a 30 ohm, 50 ohm, 75 ohm and 100 ohm microstrip line on an alu- minum substrate that has a relative dielectric constant of 9.9.

3.15 What is the impedance of a mircrostrip line that is 1.4 mm wide on a substrate that is 0.6 mm thick and has a relative dielectric con- stant of 2.2?

3.8 References

1. Bahl I. J. and D. K. Trivedi. "A Designer's Guide to Microstrip Line." *Microwaves,* (May 1977): 174-182.

2. Edwards, C. *Foundations for Microstrip Circuit Design.* New York: John Wiley and Sons, 1981.

3. Gunston, M. A. R. *Microwave Transmission Line Impedance Data.* 2nd ed. Atlanta: Noble Publishing, 1997.

4. Gupta, K. C., Ramesh Garg, Inder Bahl, and Prakash Bhartia. *Microstrip Lines and Slotlines.* Norwood: Artech House, 1996.

5. Smith, Phillip. *Electronic Applications of the Smith Chart.* Atlanta: Noble Publishing, 1995.

4

S-Parameter Circuit Analysis

4.1   Introduction

S-parameters and transmission lines are the main tools utilized in microwave amplifier design. Before we can design our own circuits, however, we need to study some methods of analysis. In this chapter, we will use mathematical and graphical tools to analyze circuits. These circuits will be networks composed of one or more elements. The methods shown in this chapter assume that the S-parameters of a one or two-port have been found or measured. We will concentrate on analyzing the circuit response of a network and groups of interconnected networks. The tools presented in this chapter as well as in Chapter 3 lay the groundwork for the following chapters on synthesizing microwave circuits and amplifiers.

Section 4.2 introduces complex mapping functions. Complex mapping functions are often needed when using transmission lines and S-parameters and are essential in identifying zones of acceptable or unacceptable impedance loads that can be placed on a transistor. As a casual observer, you can always turn to formulas in this book or other books without having to understand their origin. However, behind the scenes, it is important to understand the functions, properties, and theorems used to develop these formulas. Mapping functions are a way to analyze how a complex function affects a complex variable. For example, in this chapter, we will be discussing how an impedance is changed by connecting various circuits, such as a transistor, to it. One way to determine this is to investigate how the impedance plane is distorted, or mapped, onto another complex plane. We then can look at plots of this mapping to analyze a particular circuit or design. We will be very interested in areas in the mapping where there is negative resistance or a reflection coefficient greater than 1. When this is the case, oscillations may occur in the amplifier.

Section 4.3 shows how the mapping functions apply to S-parameters in the analysis of two-ports. Many of the equations encountered in the analysis and design of linear networks take the same form. In Example 3.8, we saw how a terminated two-port is converted to a one-port by algebraic reduction of the network's linear equations. In Section 4.3, we will see how mapping functions are used to map an impedance terminating one port onto the input impedance of another port.

Section 4.4 describes a powerful method of circuit analysis called signal flow graphing. This analysis technique begins by graphing a network as a collection of signal paths and nodes. By following a few simple rules, this flow graph can be reduced, one node or signal line at a time, to a simple network. Signal flow graphing is a quick and easy way to analyze circuits and networks. Once you learn four rules, you can analyze large networks without having to solve a series of simultaneous equations.

Section 4.5 gives the S-parameters of some common circuit elements and a general method of S-parameter analysis appropriate for computer-aided design. Transmission lines, series impedances, shunt impedances and other simple networks will be listed and referred to in later chapters. We will then briefly describe a general technique for analyzing an arbitrary collection of interconnected multi-ports. This method can be computationally intensive and is well suited for a computer-aided design system.

4.2   Complex Mapping Functions

In this section, we will describe and define the many mapping functions used to analyze and design microwave circuits and discover what may and may not be an optimum amplifier design. For example, sometimes it is not possible to design an amplifier with the highest possible gain because the circuit may oscillate. Also, it might be impossible to optimize the gain and noise performance of a transistor amplifier. Thus, we need to make some qualitative and subjective judgments as to what is acceptable. Mapping functions will be used to segment and identify regions on the Smith chart to help in making these decisions.

When working with valued functions of a real variable, we can study the functions' properties by graphing them. For example, the equation for a circle centered at $x = 0$ and $y = 0$ is

$$x^2 + y^2 = r^2$$

4.1

Equation 4.1 can be conveniently plotted on a x-y grid showing the function. However, when working with complex numbers, we do not have a similarly convenient way for studying the relation between dependent and independent variables. These variables exist as pairs of numbers on a plane rather than on a line. We can compare the two complex planes for the independent and dependent variables. Some information about a function can be gained by noticing how certain collections of points, such as lines and circles, are transformed by the function. This is called *mapping*, a process of transforming from one complex plane to another with a function and comparing the result with the original. Let us take a simple example of a mapping from the complex s-plane to another complex plane (w) using the transformation

$$w = s + 1 \qquad\qquad 4.2$$

All points in the s-plane are transformed through the function and get mapped on the w-plane one unit to the right. This is called a *translation* and is characterized by the general formula

$$w = s + O \qquad\qquad 4.3$$

where O can be any complex number.

We will frequently use three other simple transformations: *scaling, rotation*, and *reflection*. Scaling describes an enlargement or reduction in the size of the s-plane when mapping to the w-plane, as described by the function

$$w = x * s \qquad\qquad 4.4$$

where x is real. Rotation is a mapping that results in the s-plane being rotated through the function onto the w-plane. A rotation is characterized by the function

$$w = r * s \qquad\qquad 4.5$$

where r is a complex number with a magnitude equal to 1. We have seen rotation in Chapter 3 when we studied the input reflection coefficient of a terminated lossless transmission line. Finally, reflection describes the process of flipping the plane, or creating a mirror image of the s-plane in the w-plane, and is accomplished by conjugating s.

These simple mapping functions are used when designing matching circuits on the input and output of transistor amplifiers. Determining which kind of circuit to design requires more elaborate mapping functions. In Example 3.8 we saw how a termination on Port 2 of a two-port appears at Port 1. Often there will be a reflection coefficient that has to be mapped through a two-port network. These mapping functions are in the form of an inverse, or a bilinear equation. Consider the inverse function

$$w = \frac{1}{s}$$

4.6

The simplest way to map this function is to work in polar coordinates, but we will learn more when we use rectangular coordinates.

$$w = u + iv = \frac{1}{x + jy}$$

4.7

$$(u + iv)(x + iy) = 1$$

4.8

Next, we equate the real and imaginary parts.

$$ux - vy = 1 \quad \text{and} \quad vx + uy = 0$$

4.9

We first eliminate x from the equations and then y. The result is the family of circles.

$$\left(v + \frac{1}{2y}\right)^2 + u^2 = \frac{1}{4y^2}$$

4.10

$$v^{2+} \left(u + \frac{1}{2x}\right)^2 = \frac{1}{4x^2}$$

4.11

The two cases in which $x = 0$ and $y = 0$ cannot be evaluated by the above equations. But we find that these two conditions correspond to the two axes at $u = 0$ and $v = 0$ (circles of infinite radius and a center at infinity). Figure 4-1 shows how the Cartesian grid in the s-plane is mapped to the w-plane.

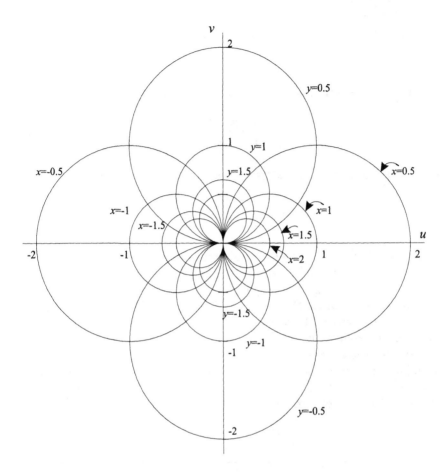

Figure 4-1 *Mapping of s to w =1/s.*

Another common transformation is the linear fractional transformation,

$$w = \frac{as+b}{cs+d}$$

4.12

which can also be written as

$$csw + bs + dw + b = 0$$

4.13

and is called the *bilinear equation*. When $c = 0$, the transformation is reduced to a scaling and a translation.

$$w = \frac{a}{b}s + \frac{b}{d} \tag{4.14}$$

When c is not equal to zero, the bilinear equation can be written as

$$w = \frac{a}{c} + \frac{bc - ad}{c}\left(\frac{1}{cs + d}\right) \tag{4.15}$$

which is a translation, a scaling, and a rotation of the function

$$\frac{1}{cs + d} \tag{4.16}$$

as long as $(bc - ad)$ does not equal zero. Otherwise, the entire s-plane is mapped to the point a/c in the w-plane. This is equivalent to the transformation $1/s$ with a scaling, rotation, and translation of the s-plane before inversion. In all these cases, there is a one-to-one mapping of points from the s-plane to the w-plane. This is an important property that will ensure that the result of this analysis using these functions is unique.

   Example 4.1: Map a line from the s-plane to the w-plane using the transformation $w = 1/s$. The equation for a line in the complex plane is $s = x + j(mx + b)$, where m and b are real numbers. We can express the line as a function of x in the w-plane.

$$w = (u + jv)(x + j(mx + b)) = 1$$

$$w = ux - v(mx + b) + j(vx + u(mx + b))$$

The function is separated into real and imaginary parts.

$$x(u - vm) + vb = 1$$

$$x(v + um) + ub = 0$$

We then solve for x.

$$x = \frac{-ub}{v + um}$$

$$x = \frac{1 + vb}{u - vm}$$

$$\left(u = \frac{m}{26}\right)^2 + \left(v = \frac{1}{26}\right)^2 = \frac{m^2+1}{4b^2}$$

One interesting property of the bilinear transform is that a circle in the s-plane becomes mapped to a circle in the w-plane. We will frequently use this property in the design of microwave amplifiers and therefore need to develop a general formula that describes how to map circles through the bilinear transform. A circle centered at $O = (t, s)$ with a radius of R can be described in polar form as

$$s = R\,e^{j\theta} + O \qquad\qquad 4.17$$

where θ is a real number and describes the angle around the circle. Starting with Equation 4.16, a circle in the s-plane becomes

$$w = \frac{1}{Rce^{j\theta} + cO + d} \qquad\qquad 4.18$$

$$= \frac{Rce^{-j\theta} - (cO+d)^*}{|R|^2|c|^2 - |cO+d|^2} \qquad\qquad 4.19$$

This describes a circle in the w-plane with a center at

$$-\frac{(cO+d)^*}{|R|^2|c|^2 - |cO+d|^2} \qquad\qquad 4.20$$

and a radius of

$$\frac{|R||c|}{|R|^2|c|^2 - |cO+d|^2} \qquad\qquad 4.21$$

Finally, the circles in Equation 4.19 are scaled and rotated by $dc - ad/c$ and translated by a/c. After some algebra, the final mapping of the circle is centered at

$$Q = \frac{|R|^2 ac^* - (b + aO)(cO + d)^*}{|R|^2|c|^2 - |cO + d|^2} \qquad\qquad 4.22$$

and has a radius of

$$P = \frac{|R||a(d+cO) - c(b+aO)|}{|R|^2 |c|^2 - |cO+d|^2} \qquad 4.23$$

A special and very useful case of the above is the mapping of circles centered on the origin of the s-plane. Using Equation 4.22 with $O = 0$, the circle on the w-plane will be centered at

$$Q = \frac{|R|^2 ac^* - db^*}{|R|^2 |c|^2 - |d|^2} \qquad 4.24$$

and will have a radius of

$$P = \frac{|R||ad - bc|}{|R|^2 |c|^2 - |d|^2} \qquad 4.25$$

We will use these two equations when we graph the properties of a microwave amplifier on a Smith chart. Using Equations 4.24 and 4.25, we will derive equations to map a zone's stability, constant gain, and noise figure for an amplifier.

Example 4.2: Map the $|s| = 1$ circle on the w-plane using the bilinear transform in Equation 4.12 with $a = 1 + j3$, $b = 3 + j2$, $c = 2 - j4$, and $d = j6$. The $|s| = 1$ circle is centered at $s = (0, 0)$ and has a radius of 1. In this case, the center of the circle in the w-plane is given by Equation 4.24 as

$$Q = \frac{(1+j3)(2+j4) - (3+j2)(-j6)}{(20) - (36)}$$

$$Q = -\frac{(-10+j10) - (j18+12)}{16}$$

$$Q = -\frac{(2+j8)}{16} = -\frac{1}{8} - j\frac{1}{2}$$

The radius is found by using Equation 4.25.

$$P = \frac{|(1+j3)(j6) - (3+j2)(2-j4)|}{(20) - (36)}$$

$$P = \frac{|(j6 - 18) - (14 + j4)|}{(20) - (36)}$$

$$P = \frac{|(-32 + j10)|}{16} = \frac{\sqrt{1124}}{16} = \frac{\sqrt{281}}{8}$$

We have only considered mapping the bilinear equation from one complex plane to another. Although there are many other useful mapping functions, only the bilinear equation is needed in most design situations. The next section will show how these mapping functions are used as S-parameter design tools.

4.3   Mapping S-Parameters in Circuit Analysis

The previous section detailed some complex functions and showed how they are mapped. We will now apply this to S-parameter circuit analysis. The Z-, Y-, chain, and S-parameter representation for circuits have their basis in linear algebra. By understanding a few simple relations, the mechanics of circuit analysis can be accomplished.

Chapter 3 introduced the Smith chart and demonstrated how a load impedance is mapped on it. This is a mapping of a one-port from one complex plane (impedance) to another (reflection coefficient). The reflection coefficient is the ratio of the reflected-to-incident wave, or S_{11}. Example 3.8 showed how to use linear algebra to calculate the reflection coefficient at the input of a two-port when the output is terminated. This section will show how to map these functions. We will use these mappings when synthesizing amplifier circuits.

Consider the circuit shown in Figure 4-2, where a two-port is terminated on Port 2 with a load impedance that has a reflection coefficient of Γ_L. We have demonstrated that the two-port is characterized by the matrix equations

$$\begin{bmatrix} a_1 \\ a_2 \end{bmatrix} = \begin{bmatrix} S_{11} & S_{21} \\ S_{12} & S_{22} \end{bmatrix} \begin{bmatrix} b_1 \\ b_2 \end{bmatrix} \qquad 4.26$$

and the one-port, which is defined as

$$\Gamma_L = \frac{b}{a} \qquad 4.27$$

We have already seen in Example 3.8 that when $a = b_2$ and $b = a_2$, the input reflection coefficient is equal to

$$\Gamma_i = S_{11} + \frac{S_{21}S_{12}\Gamma_L}{(1 - S_{22}\Gamma_L)} = \frac{S_{11} - \Delta\Gamma_L}{(1 - S_{22}\Gamma_L)} \qquad 4.28$$

where $\Delta = S_{11}S_{22} - S_{12}S_{21}$.

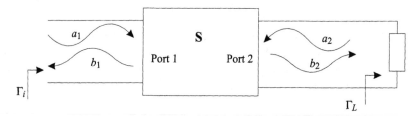

Figure 4-2 *A two-port terminated by a load.*

This equation appears in the form of a bilinear equation describing the input reflection coefficient as a function of the load reflection coefficient. We can map the Γ_L plane onto the input reflection coefficient plane using the mapping functions derived in the last section. The circle of constant $|\Gamma_L|$ is centered at the origin, so Equations 4.24 and 4.25 are used to map the circles in the Γ_i plane. These circles are centered at

$$Q = \frac{|\Gamma_L|^2 \Delta S_{22}^* - S_{11}}{|\Gamma_L|^2 |S_{22}|^2 - 1} \qquad 4.29$$

and have a radius of

$$P = \frac{|\Gamma_L||S_{21}S_{12}|}{|\Gamma_L|^2 |S_{22}|^2 - 1} \qquad 4.30$$

If we hold the magnitude of the load constant and adjust the phase of the load impedance, the input reflection coefficient will lie somewhere on that circle. An example of one such mapping is shown in Figure 4-3. Three circles of constant reflection coefficient are shown. These circles are formed in the input reflection coefficient plane by a load with a reflection coefficient of 0.8, 0.6 or 0.4.

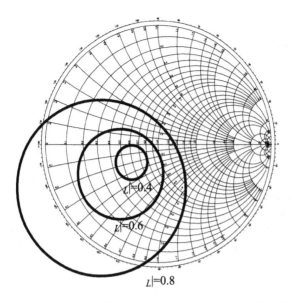

$_L|$=0.8

Figure 4-3 *An example of a mapping of three load Γ_L through a two-port network onto the input reflection coefficient plane.*

In later chapters, we will find other useful mappings that describe constant gain and power circles. These will map the input reflection coefficient through the two-port onto the output reflection coefficient plane. The stability of a transistor amplifier is guaranteed when the reflection coefficient is less than 1. When we map the input and output reflection coefficients onto the source and load impedance plane, we will be able to show which source and load impedances might cause the amplifier to be unstable.

Three-port devices are frequently used in microwave engineering. They mainly take the form of transistors that are connected to the circuit as a two-port. It will be useful to describe a transistor as a three-port when designing an amplifier with a series-feedback element on the emitter or drain. We may need transistor circuits with the base, gate, collector, or drain grounded instead of a grounded emitter or source terminal. These configurations are useful when designing oscillators and frequency mixers. If we consider a transistor as a three-port, as shown in Figure 4-4, the S-parameters of this network are

$$\mathbf{S} = \begin{bmatrix} S_{11} & S_{21} & S_{31} \\ S_{12} & S_{22} & S_{32} \\ S_{13} & S_{23} & S_{33} \end{bmatrix} \qquad 4.31$$

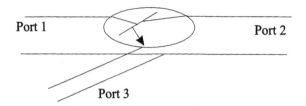

Figure 4-4 *A transistor connected as a three port.*

The transistor S-parameters are almost always given by two-port parameters measured with the source or emitter grounded, as shown in Figure 4-5. The two-port S-parameters for the common emitter or source transistor circuit are obtained by grounding port three of the transistor three-port. A short has a reflection coefficient of –1, which means

$$a_3 = -b_3 \qquad\qquad 4.32$$

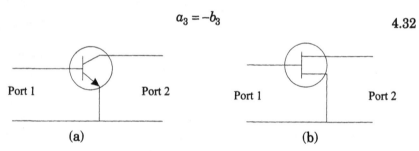

$$\text{(a)} \qquad\qquad \text{(b)}$$

Figure 4-5 *A transistor connected in a common emitter (a) or common source (b) configuration, as most data sheets would give the S-parameters of the device.*

We then use Equation 4.32 in the linear equation for a three-port (Equation 4.31)

$$\begin{bmatrix} b_1 \\ b_2 \\ b_3 \end{bmatrix} = \begin{bmatrix} S'_{11} & S'_{12} & S'_{13} \\ S'_{21} & S'_{22} & S'_{23} \\ S'_{31} & S'_{32} & S'_{33} \end{bmatrix} \begin{bmatrix} a_1 \\ a_2 \\ -b_3 \end{bmatrix} \qquad\qquad 4.33$$

and reduce the order for the linear equation by solving for b_3.

$$b_1 = S'_{11} a_1 + S'_{12} a_2 - S'_{13} b_3 \qquad\qquad 4.34$$

$$b_2 = S'_{12} a_1 + S'_{22} a_2 - S'_{23} b_3 \qquad\qquad 4.35$$

$$b_3 = S'_{31} a_1 + S'_{32} a_2 - S'_{33} b_3 \qquad\qquad 4.36$$

$$b_3 = \frac{S'_{31}a_1 + S'_{32}a_2}{1+S'_{33}}$$ 4.37

We use Equation 4.37 and Equations 4.34 and 4.35 to yield the two-port S-parameters of the three-port with port three shorted to ground.

$$S_{11} = S'_{11} - \frac{S'_{13}S'_{31}}{1+S'_{33}}$$ 4.38

$$S_{12} = S'_{12} - \frac{S'_{13}S'_{32}}{1+S'_{33}}$$ 4.39

$$S_{21} = S'_{21} - \frac{S'_{23}S'_{31}}{1+S'_{33}}$$ 4.40

$$S_{22} = S'_{22} - \frac{S'_{23}S'_{32}}{1+S'_{33}}$$ 4.41

If we had solved for the two-port S-parameters by terminating port three in any reflection coefficient (Γ_3), Equations 4.38 through 4.41 would be bilinear equations with Γ_3 appearing in the numerator and denominator of every equation. We can reverse the above process to find the three-port S-parameter as a function of the two-port parameters. When the two-port S-parameters are given with one terminal grounded, the three-port S-parameters can be found to be

$$S'_{33} = \frac{S_{11} + S_{12} + S_{21} + S_{22}}{4 - S_{11} - S_{12} - S_{21} - S_{22}}$$ 4.42

$$S'_{32} = \frac{1+S'_{33}}{2}\left(1 - S_{12} - S_{22}\right)$$ 4.43

$$S'_{23} = \frac{1+S'_{33}}{2}\left(1 - S_{21} - S_{22}\right)$$ 4.44

$$S'_{22} = S_{22} + \frac{S'_{23}S'_{32}}{1+S'_{33}}$$ 4.45

$$S'_{31} = 1 - S'_{32} - S'_{33}$$ 4.46

$$S'_{13} = 1 - S'_{23} - S'_{33}$$ 4.47

$$S'_{12} = 1 - S'_{32} - S'_{22}$$ 4.48

$$S'_{21} = 1 - S'_{23} - S'_{22} \qquad\qquad 4.49$$

$$S'_{11} = 1 - S'_{21} - S'_{31} \qquad\qquad 4.50$$

Example 4.3: Map the circle of constant source reflection coefficient through a two-port onto the output reflection coefficient plane as shown in Figure 4-6. The source reflection coefficient Γ_s is the ratio of waves exiting the source to waves entering the source.

$$\Gamma_s = \frac{a_1}{b_1}$$

$$a_1 = \Gamma_s b_1$$

Then, using the linear equations for the two-port, we solve for $\Gamma_o = b_2/a_2$.

$$b_1 = S_{11}\Gamma_s b_1 + S_{12}a_2$$

$$b_1 = \frac{S_{12}a_2}{1 - S_{11}\Gamma_s}$$

$$b_2 = S_{21}\Gamma_s \frac{S_{12}a_2}{1 - S_{11}\Gamma_s} + S_{22}a_2$$

$$\Gamma_o = S_{22} + \frac{S_{21}S_{12}\Gamma_s}{1 - S_{11}\Gamma_s} = \frac{S_{22} - \Delta\Gamma_s}{1 - S_{11}\Gamma_s}$$

where $\Delta = S_{11}S_{22} - S_{12}S_{21}$. Circles of constant $|\Gamma_s|$ are centered at the origin in the Γ_s plane. We can use the mapping Equations 4.24 and 4.25. The center of the $|\Gamma_s|$ circle will appear on the Γ_o plane centered at

$$b_2 = S_{21}\Gamma_s \frac{S_{12}a_2}{1 - S_{11}\Gamma_s} + S_{22}a_2$$

and will have a radius of

$$P = \frac{|\Gamma_s||S_{21}S_{12}|}{|\Gamma_s|^2|S_{11}|^2 - 1}$$

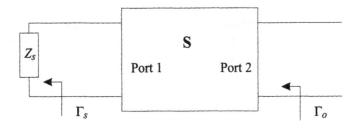

Figure 4-6 *A two-port with the input loaded by a source impedance.*

The bilinear equation appears frequently in network analysis and design problems and is part of Z-, Y-, and S-parameter analysis situations. We will need to know how certain values of source or load impedance appear on the other port of a transistor amplifier. By using the mapping functions, we can plot these zones on a Smith chart, which provides a convenient way to analyze the effect of loading one of the ports of a network.

4.4   Signal Flow Graphs

In the previous two sections, we have shown how to analyze S-parameter networks by solving a set of linear equations representing these networks. The input reflection coefficient of an output-terminated two-port was found by solving a set of equations. For example, in Section 4.3 the S-parameters of a three-port were converted to the S-parameters of a two-port by terminating port three and solving the set of linear equations. The new two-port S-parameters were found by reducing the three-by-three matrix to a two-by-two matrix.

In this section, we will introduce a method of circuit analysis using signal flow graphs. With the use of a few simple rules, the signal flow method can be used to expedite circuit analysis and help in visualizing the process. Signal flow graphs are particularly well-suited for the analysis of specific circuit problems and can be useful when analyzing interconnected networks. The calibration of test equipment — after errors have been measured in the test equipment and removed from the measurement of the test device — is a common example. The process is called *de-embedding* the imperfections of the test equipment from the measurements.

We will begin by analyzing a simple network. Figure 4-7 shows an idealized feedback network and its signal flow diagram with arbitrary transfer functions. Each function is represented by a simple circuit and the corresponding signal flow diagram. A signal flow graph has two properties. First, a network is represented by signal paths that are connected by nodes.

Second, signal paths are indicated by the arrow lines and a transfer function that is associated with them. Signals travel only in one direction along a line. (The direction is indicated by the arrow.) Signals flow into or out of nodes that split or sum these signals.

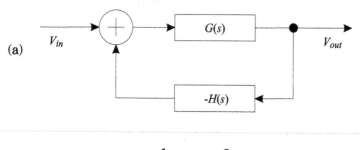

(a)

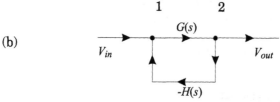

(b)

Figure 4-7 *A simple feedback network and its corresponding signal flow graph.*

The signal flow diagram in Figure 4-7 shows two nodes and four signal paths. The input signal enters the diagram along the path labeled V_{in}. The output is represented by the signal path labeled V_{out}. The feedback network takes signals from node 2 and sends them to node 1. Using the circuit diagram in Figure 4-7(a), we see that the output signal is the sum of the input and feedback signals that have been sent through $G(s)$.

$$V_{out} = G(s)\left(V_{in} + V_f\right) \qquad 4.51$$

The feedback signal is

$$V_f = -G(s)H(s)\left(V_{in} + V_f\right) \qquad 4.52$$

Solving for V_f and substituting it into Equation 4.51, the transfer function is

$$\frac{V_{out}}{V_{in}} = \frac{G(s)}{1 - G(s)H(s)} \qquad 4.53$$

We can now reduce the graph of Figure 4-7(b) to the same answer that was found in the above equation.

There are two basic methods of analyzing signal flow graphs. Mason's nontouching loop rule requires finding all unique signal paths through the network. However, it is easy to overlook a path when using Mason's method. The second method, known as the signal flow graphing technique, calls for the reduction of the signal flow graph until there is only one path from input to output.

In the process of reducing the signal flow graph, we will need to combine paths and nodes. There are four simple rules that dictate how to combine signal paths. The first rule states that when two paths are in series and have the same direction, they can be combined by multiplying the transfer functions of the two paths. It is important to make sure that the two paths have the same direction; otherwise they cannot be combined. Figure 4-8(a) shows how two paths, connected in series, are combined into one path. This action eliminates a node in the process.

The second rule states that when two paths are in parallel and have the same direction, they can be combined by adding the two transfer functions. As in the first rule, both paths must have the same direction. Figure 4-8(b) shows two parallel paths and their combination.

The third rule prescribes how to eliminate self-loops. A self-loop is a signal path that originates and terminates at the same node, as shown in Figure 4-8(c). A self-loop can be eliminated if all incoming signals are divided by 1 minus the self-loop transfer function. Only the incoming signals are affected by this process.

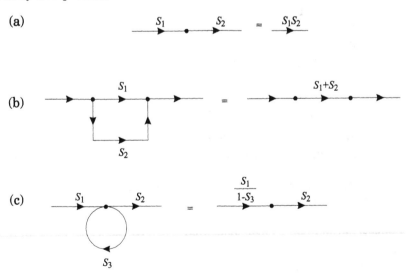

Figure 4-8 *The three ways to combine nodes and paths in a signal flow graph: (a) two series paths, (b) two parallel paths, and (c) a self-loop node.*

These three rules explain how to reduce a signal flow graph one signal or node at a time. A fourth rule dictates how to split a complicated node into one or more individual nodes, which is necessary in order for the signal paths to be reduced one at a time. According to this rule, a node can be split as long as each input-output combination is represented once, and only once, in the new graph. If the node includes a self-loop, the self-loop must be duplicated in all the split nodes. Figure 4-9 shows how a node with two incoming signals, three outgoing signals, and a self-loop can be split into two nodes. The possible combinations of input and output paths are

$$S_1 \text{ to } S_3$$
$$S_2 \text{ to } S_3$$
$$S_1 \text{ to } S_4$$
$$S_2 \text{ to } S_4$$
$$S_1 \text{ to } S_5$$
$$S_2 \text{ to } S_5$$

All of these six paths are represented in the two split nodes, along with a duplicate of the self-loop on both of the new nodes.

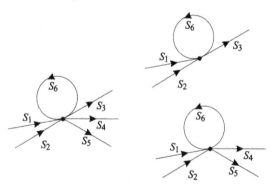

Figure 4-9 *A signal node is split into two nodes by conserving each combination of input and output paths and duplicating the self-loop in each new node.*

We can now reduce the signal flow diagram in Figure 4-7(b). Although there are two parallel paths, $H(s)$ and $G(s)$, they are not going in the same direction, so the second rule cannot be used. The signal diagram cannot be reduced without splitting a node. Figure 4-10(b) shows node 1 split into two nodes. Notice that there are two combinations of input and output signals represented in both Figure 4-10(a) and 4-10(b): V_{in} to $G(s)$ and $-H(s)$ to $G(s)$. The series combination of $G(s)$ and $-H(s)$ eliminates node 1' and creates a self-loop, as shown in Figure 4-10(c). At the end, the self-loop is absorbed into the $G(s)$ path, yielding the same result as in Equation 4.53

above. We have now reduced the signal flow graph to a single equation. Figure 4-11 shows the signal flow graph for a two-port network. The four nodes and four paths with their S-parameters describe the network.

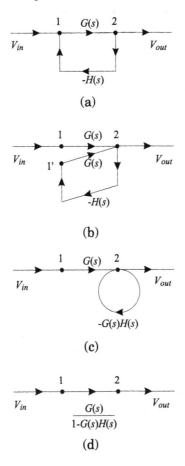

(a)

(b)

(c)

(d)

Figure 4-10 *The process of reducing the signal flow graph in Figure 4-7(b) to a single path by (b) splitting a node, (c) series combination, and (d), a self-loop reduction.*

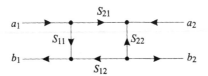

Figure 4-11 *The signal flow graph for a two-port network described with S-parameters.*

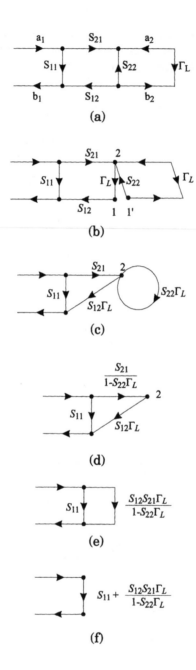

Figure 4-12 *A terminated two-port signal flow graph and how it is collapsed to a one-port using the four rules stated in this section.*

Example 4.4: Find the expression for the input reflection coefficient of a two-port with a termination on the output. Figure 4-12(a) shows the signal flow graph of an output-terminated two-port. Starting at the output load, there are two parallel paths which cannot be combined because they are pointing in opposite directions. Thus, node 1 is split into two nodes, as shown in Figure 4-12(b).

The series connection of S_{12} and Γ_L is combined into one path [see Figure 4-12(c)]. Also, the series combination of S_{22} and Γ_L makes a self-loop. The self-loop is absorbed into the signal path-entering node 2, as shown in Figure 4-12(d). Further, two series paths are combined in Figure 4-12(e). Two parallel paths pointing in the same direction can be combined [Figure 4-12(f)].

Signal flow graphing is well-suited for medium sized networks. It is especially useful for analyzing the effect of imperfect test equipment on a measurement and removing that effect to yield a more accurate measurement. There are only four rules for reducing a signal flow graph. First, we multiply the two transfer functions combines series paths that flow in the same direction. Second, adding the two transfer functions can combine parallel paths that flow in the same direction. Third, a lone signal loop, or self-loop, is absorbed into the signal paths entering the node. And fourth, node splitting is used to reduce the complexity of a signal system entering or leaving a node. Sometimes it is much easier to analyze circuits by condensing signal graphs than by solving algebraically. Figure 4-13 shows a signal flow diagram of a two-port S-parameter test where the test equipment is not perfect. Error terms e_{ij} account for inaccuracies of the test equipment, cables, and connectors.

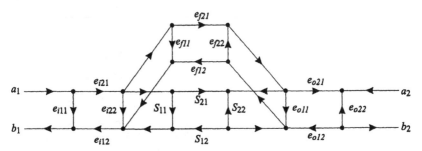

Figure 4-13 *The signal flow graph for a two-port S-parameter test setup where 12 error terms are accounted for in the measurement.*

4.5 General Matrix S-Parameter Solutions

A few common circuits that will be used when designing microwave circuits are shown in Figure 4-14 along with the S-parameter matrix representation of each circuit. The transmission line in Figure 4-14(a) has been given in Chapter 3, where Z is the transmission line impedance, and θ is the electrical length in radians. The shunt admittance in Figure 4-14(b) and series impedance in Figure 4-14(c) have been shown in Chapter 2. The π-network and T-network in Figure 4-14(d) and 4-14(e) are common impedance-matching elements. The open and shorted transmission line stubs shown in Figure 4-14(e) and 4-14(f) are used frequently in amplifier-matching circuits.

$$Z, \theta$$

$$S = \begin{bmatrix} \dfrac{\Gamma(1-e^{-2j\theta})}{1-\Gamma^2 e^{-2j\theta}} & \dfrac{(1-\Gamma^2)e^{-2j\theta}}{1-\Gamma^2 e^{-2j\theta}} \\ \dfrac{(1-\Gamma^2)e^{-2j\theta}}{1-\Gamma^2 e^{-2j\theta}} & \dfrac{\Gamma(1-e^{-2j\theta})}{1-\Gamma^2 e^{-2j\theta}} \end{bmatrix}$$

$$\Gamma = (Z - Z_o)/(Z + Z_o)$$

(a)

$$y = Y/Y_o$$

$$S = \begin{bmatrix} \dfrac{-y}{(y+2)} & \dfrac{2}{(y+2)} \\ \dfrac{2}{(y+2)} & \dfrac{-y}{(y+2)} \end{bmatrix}$$

(b)

$$z = Z/Z_o$$

$$S = \begin{bmatrix} \dfrac{-y}{(y+2)} & \dfrac{2}{(y+2)} \\ \dfrac{2}{(y+2)} & \dfrac{-y}{(y+2)} \end{bmatrix}$$

(c)

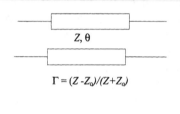

$$y_3 = Y_3/Y_o$$

$$y_1 = Y_1/Y_o \qquad y_2 = Y_2/Y_o$$

$$\Delta_y = Y_1Y_2 + Y_1Y_3 + Y_2Y_3$$

$$S = \begin{bmatrix} \dfrac{1 - y_1 + y_2 - \Delta_y}{D_y} & \dfrac{-2y_3}{D_y} \\ \dfrac{-2y_3}{D_y} & \dfrac{1 + y_1 - y_2 - \Delta_y}{D_y} \end{bmatrix}$$

$$D_y = 1 + y_1 + y_2 + 2y_3 + \Delta_y$$

(d)

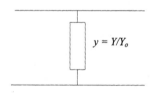

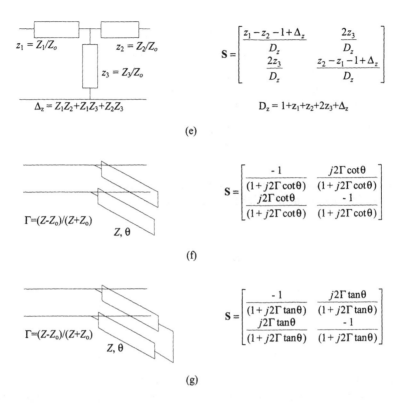

Figure 4-14 *Common circuit elements and their S-parameters: (a) a series transmission line, (b) a shunt admittance, (c) a series impedance, (d) a π-network, (e) a T-network, (f) an open shunt transmission line, (g) and a shorted shunt transmission line.*

Any arbitrary combination of interconnected networks can be analyzed algebraically, an approach that is well-suited for a computer-aided design process. Figure 4-15 shows a network with an arbitrary number of n-ports. Let us assume there are m number of subnetworks connected inside the main network. The network is expressed in matrix form as

$$\mathbf{b} = \mathbf{S}\mathbf{a} \qquad 4.54$$

where $\mathbf{a}$ and $\mathbf{b}$ are column vectors composed of all the column vectors from the subnetworks

$$
\mathbf{a} = \begin{bmatrix} a^{(1)} \\ a^{(2)} \\ \vdots \\ a^{(k)} \\ \vdots \\ a^{(m)} \end{bmatrix} \qquad \mathbf{b} = \begin{bmatrix} b^{(1)} \\ b^{(2)} \\ \vdots \\ b^{(k)} \\ \vdots \\ b^{(m)} \end{bmatrix} \qquad \text{4.55}
$$

and the S-parameter matrix includes the S-parameters of the subnetworks on the diagonal of a (mn) by (mn) matrix.

$$
\mathbf{S} = \begin{bmatrix} S^{(1)} & 0 & \cdots & 0 & \cdots & 0 \\ 0 & S^{(2)} & & & & 0 \\ \vdots & & \ddots & & & \vdots \\ 0 & & & S^{(k)} & & 0 \\ \vdots & & & & \ddots & \vdots \\ 0 & 0 & \cdots & 0 & \cdots & S^{(m)} \end{bmatrix} \qquad \text{4.56}
$$

Vectors $\mathbf{a}$ and $\mathbf{b}$ are constrained by the way the subnetworks are interconnected. Next, we relate the traveling waves $a^{(k)}{}_i$ and $b^{(g)}{}_j$ wherever two of the subnetworks are interconnected

$$
\mathbf{b} = \mathbf{\Gamma a} \qquad \text{4.57}
$$

where Γ is the connection matrix. The elements of Γ are equal to zero except where two or more nodes are connected. Wherever ports are connected, the elements of Γ are the S-parameters of the interconnect. When the ports are perfectly matched, Γ is equal to the identity matrix. If it is not the case, Γ accounts for the reflections between ports inside the network. For example, when two mismatched ports are connected, the elements of the connection matrix are

$$
\frac{1}{z_i + z_j} \begin{bmatrix} z_i - z_j{}^* & 2\sqrt{\mathrm{Re}(z_i)\,\mathrm{Re}(z_j)} \\ 2\sqrt{\mathrm{Re}(z_i)\,\mathrm{Re}(z_j)} & z_i - z_j{}^* \end{bmatrix} \qquad \text{4.58}
$$

After combining Equations 4.54 and 4.58, the network equation becomes

$$\mathbf{b} = (\mathbf{S} - \mathbf{G})\mathbf{a} \qquad 4.59$$

The complexity of the equations makes this analysis technique more suitable for computer-aided design programs. As the number of subnetworks grows and the complexity of the interconnects increases, it becomes a tedious task to find the S-parameters of the composite network.

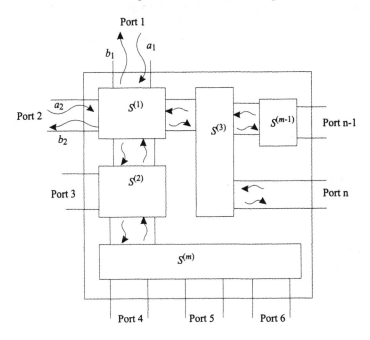

Figure 4-15 *An arbitrary network of mn-ports.*

4.6   Summary

We have shown a few complex mapping functions including the bilinear equation. The bilinear equation occurs every time an n-port is converted to an (n-1)-port by terminating one port. Every S-parameter in the (n-1)-port appears in the form of a bilinear equation. For example, when a two-port is terminated on the output, the input reflection coefficient of the two-port is a bilinear equation. The mapping function results in a one-to-one mapping of the termination onto the input impedance on the other port. This property is used to find constant gain circles, stability circles, and constant noise figure circles in future chapters. Equations 4.24 and 4.25 are the only equations needed to find the location of these circles.

Signal flow graphs are a simple way of analyzing S-parameter circuits and networks. By using four simple rules, a S-parameter circuit can be reduced to a simple circuit. Parallel signal paths flowing in the same direction can be combined by adding the two transfer functions. Multiplying the two transfer functions can combine series signal paths. A self-loop can be absorbed into the network by dividing all the incoming signal path transfer functions by 1 minus the loop transfer function. The last rule states that a node can be split, providing that all possible combinations of input and output signal paths are accounted for exactly once.

The last section showed the S-parameters of some common circuits. Series and shunt transmission lines are used as circuit elements in microwave amplifiers. The S-parameters of these and a few other elements were listed. A general method of S-parameter circuit analysis was briefly introduced. This technique is applicable to any number of arbitrarily connected multi-ports. However, the complicated algebra involved in the analysis makes this method more suitable to computer-aided design applications.

4.7   Problems

4.1   Map the $x = 3$ line on the w-plane, using the transformation $w = 1/s$.

4.2   Map the $y = 5$ line on the w-plane, using the transformation $w = (2 + j4)/s$.

4.3   Map the line $3x + 5 = y$ in the s-plane into the w-plane, using the transformation $w = 1/(3 + j4 + s)$.

4.4   A circle centered at $6 + j7$ and having a radius of 2 is transformed through the function $1/s = w$. Find the new center and radius.

4.5   A circle centered at $3 - j2$ and having a radius of 3 is transformed through the function $s/(2s + j6 - 3) = w$. Find the new center and radius.

4.6   A microwave noise source has a reflection coefficient that varies in production. The average reflection coefficient is $.2\angle45$ degrees, but the magnitude can vary $+/-.05$ at any phase. This defines a circular region in the s-plane. Find out where this region is mapped on the w-plane when the transfer function is the one used in Example 4.2.

4.7   Figure 4-16 shows a signal flow graph of a two-port measurement device that was used to measure the load Γ_L. Show how a perfect load, open and short, is sufficient to find all the error terms (e_{ij}) if $e_{12} = e_{21}$.

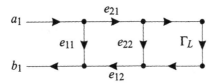

Figure 4-16 *An imperfect two-port used to measure a load impedance.*

4.8   Prove that the S-parameter matrix in Figure 4-14(b) is that of a shunt impedance.

4.9   Prove that the S-parameter matrix in Figure 4-14(c) is that of a series impedance.

5

Narrowband Circuit Synthesis

5.1 Introduction

This chapter discusses the design of impedance-matching circuits. It often is necessary to connect components or networks that have different impedances. If these components were to be indiscriminately connected, energy would be reflected between them, causing energy loss and other unwanted effects. We can use impedance-matching circuits between components in a network so that power is not "bounced around" between them. For example, many televisions come with a 75 ohm type F connector instead of 300 ohm antenna terminals. Suppose you wanted to use an antenna that has a 300 ohm impedance. If a transmission line with a characteristic impedance of 75 ohms is terminated with a 300 ohm load, the load has a reflection coefficient of

$$\Gamma = \frac{300 - 75}{300 + 75} = 0.6$$

i.e., 60% of the power will be reflected at the antenna terminals. To prevent this, we need a circuit to match the 70 ohm transmission line to the 300 ohm antenna terminals.

Impedance-matching circuits are used in two ways. The most common application is to eliminate reflections between components or networks. Most transistors are not matched to the transmission lines that connect them; hence, small circuits are designed to match the input and output impedance of the transistor to the input and output transmission lines.

Matching a transmission line to a load assures that power is efficiently extracted from the source or delivered to the load by the transistor. Loads that are not matched reflect energy, making the power transfer between these circuits inefficient. The second use for matching circuits arises when two components are purposely mismatched to achieve a specific goal. It is common practice to reflect power from a transistor to reduce amplification. For example, a broadband amplifier could be designed to have a very good match at the highest frequencies. The maximum gain of a transistor usually decreases with frequency. Therefore, the transistor needs to be well-matched at the highest frequency of interest where the natural gain of the transistor is low. At lower frequencies, the transistor is purposely mismatched to reduce its gain. If done correctly, the gain can be the same across the entire band.

In this chapter, we will show five circuit topologies for matching one impedance to another. Although the synthesis problem will not be rigorously addressed, we will show how to match impedances in the context of specific circuit topologies. Our aim is to present some common examples of circuits in order to be able to design amplifiers. Naturally, many other types of circuits that can be used. The circuits discussed in this chapter, however, are sufficient to provide an introduction. A discussion of narrowband amplifier design techniques can be found in Chapter 6.

Section 5.2 shows how to design simple matching circuits using two components. It is assumed that these components are ideal capacitors or inductors. Even though this can be an erroneous assumption at some frequencies, these circuits can be used at RF frequencies and in monolithic microwave circuits.

Section 5.3 presents the design of matching circuits using transmission lines. When beginning at a few gigahertz, capacitors and inductors will usually not function properly, and it is preferable to use small sections of transmission lines, such as microstrip lines, as the tuning elements in the matching circuits. This section will show how to design simple matching circuits using a series transmission line.

Section 5.4 and Section 5.5 introduce design matching circuits that use a simple combination of a series transmission line and an open or shorted transmission line. The input impedance of an ideal open-circuit or short-circuit transmission line is purely imaginary. These impedances simulate a capacitance or inductance. We will employ this property to design circuits using only transmission lines.

Finally, Section 5.6 discusses how to design a matching circuit using three transmission lines. Two transmission line stubs separated by a series transmission line characterize the circuit topology shown in this section. Some impedances are difficult or impossible to match using the circuits discussed in Sections 5.3, 5.4, or 5.5. Using these double-stub circuits removes some of the deficiencies of the simple circuits in the previous sections. Double-stub matching circuits are only one way of utilizing three transmission lines to synthesize a matching circuit.

5.2 Lumped-Element Lossless Matching Circuits

This section shows how to design impedance-matching circuits using a simple L circuit shown in Figures 5-1(a) and 5-1(b). The circuits in Figure 5-1 are constructed with capacitors [Figure 5-1(a)] and inductors [Figure 5-1(b)] and represent lossless circuit elements. We can use the circuit in Figure 5-1(a) to match a load $(Z_L = 1/Y_L)$ to a transmission line with a characteristic impedance of Z_o. We want to choose the shunt susceptance jB to equalize the real part of Z' and Z_o. We can then cancel the remaining imaginary part of Z', adding a series reactance jX.

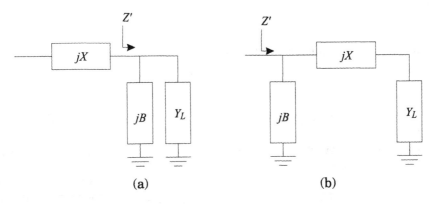

(a)   (b)

Figure 5-1 *The two L circuit topologies for matching a load impedance.*

Beginning with the circuit in Figure 5-1(a), the impedance looking at the parallel combination of the load and shunt susceptance is

$$Z' = \frac{1}{jB + Y_L} \qquad 5.1$$

$$Z'(jB + G_L + jB_L) = 1 \qquad 5.2$$

Equation 5.2 is separated into real and imaginary parts.

$$\text{Re}\{Z'\}G_L - \text{Im}\{Z'\}(B + B_L) = 1 \qquad 5.3$$

$$\mathrm{Re}\{Z'\}(B + B_L) + \mathrm{Im}\{Z'\}G_L = 0 \qquad 5.4$$

We want to force the real part of Z' to be equal to Z_o. There are two equations and two unknowns: $\mathrm{Im}\{Z'\}$ and B. We substitute Z_o for $\mathrm{Re}\{Z'\}$ in Equation 5.4 and solve for $\mathrm{Im}\{Z'\}$ first.

$$\mathrm{Im}\{Z'\} = -\frac{Z_o(B + B_L)}{G_L} \qquad 5.5$$

Inserting this into Equation 5.3,

$$Z_o(B + B_L)^2 - \left(Z_o G_L^2 - G_L\right) = 0 \qquad 5.6$$

we can solve for B

$$B = -B_L \pm \sqrt{Y_o G_L - G_L{}^2} \qquad 5.7$$

Finally, X is chosen to cancel the $\mathrm{Im}\{Z'\}$.

$$X = -\mathrm{Im}\{Z'\} \qquad 5.8$$

The circuit topology shown in Figure 5-1(a) can only be used if $Z_o G_L < 1$; otherwise the quantity inside the square root will be negative. We can see this using a Smith chart. Figure 5-2(a) shows an admittance Smith chart with a load plotted at point Y_L. The shunt susceptance (jB) can only move the load admittance along a constant conductance circle. We use the shunt element to move the load to the constant Z_o resistance line, which is shown as a dotted circle in Figure 5-2(a). Only the load admittances in the shaded area can be moved to the constant Z_o resistance circle by adding a shunt reactive element. When $\mathrm{Re}\{Y_L\} < Y_o$ [see shaded area in Figure 5-2(a)] there are two places where the constant $\mathrm{Im}\{Y_L\}$ circle intercepts the constant Z_o resistance circle. This explains the $\pm$ sign in Equation 5.7. If the load conductance is less than Y_o, the constant G_L line never intercepts the constant Z_o resistance circle.

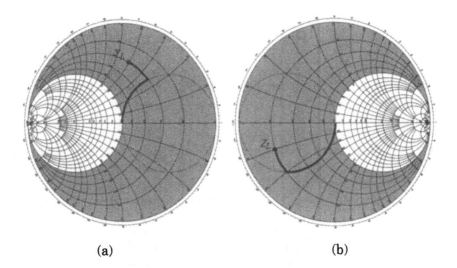

(a)	(b)

Figure 5-2 *The possible zones where the two matching circuits of Figure 5-1 can be used. The circuit in Figure 5-1(a) can be used for loads in the shaded area of Smith chart (a) above. The circuit in Figure 5-1(b) can be used to match loads in the shaded area of Smith chart (b) above.*

We can perform the same analysis using the circuit topology of Figure 5-1(b). The goal here is to choose the series impedance in a way that the admittance looking into the series combination of jX and Z_L has a real part equal to the characteristic admittance Y_o.

$$Y' = \frac{1}{jX + Z_L}$$ 5.9

$$Y'(jX + R_L + jX_L) = 1$$ 5.10

Equation 5.9 is separated into real and imaginary parts.

$$\text{Re}\{Y'\}R_L - \text{Im}\{Y'\}(X + X_L) = 1$$ 5.11

$$\text{Re}\{Y'\}(X + X_L) + \text{Im}\{Y'\}R_L = 0$$ 5.12

Again, we want to force the real part of Y' to equal Y_o. There are two equations and two unknowns: Z and $\text{Im}\{Y'\}$. We substitute Y_o for $\text{Re}\{Y'\}$ and solving for $\text{Im}\{Y\}$ and Z yields

$$\text{Im}\{Y'\} = -\frac{Y_o(X + X_L)}{R_L}$$ 5.13

and

$$X = -X_L \pm \sqrt{Z_o R_L - R_L^2} \qquad 5.14$$

The shunt element B is chosen to cancel the imaginary part of Y'.

$$B = -\text{Im}\{Y'\} \qquad 5.15$$

For the quantity inside the square root to be positive, $Y_o R_L$ must be less than 1. This area is plotted on the impedance Smith chart in Figure 5-2(b). The series reactance can only move the load impedance along constant resistance circles. Only impedances in the shaded area can be moved to the constant Y_o conductance circle shown by the dotted circle.

   Example 5.1: Design a matching circuit using the topology shown in Figure 5-1(a) or Figure 5-1(b) to match a $25 + j60$ ohm load to a 50 ohm transmission line. First, let us try the circuit in Figure 5-1(a) and test whether $Z_o G_L < 1$.

$$Y_L = \frac{1}{25 + j60} = 0.00592 - j0.0142$$

$$Z_o G_L = 0.2959$$

Since we can use the circuit in Figure 5-1(a), we begin by solving for B using Equation 5.7 and X using Equation 5.8.

$$B = 0.0142 \pm \sqrt{0.02 * 0.00592 - 0.00592^2}$$

$$B = 0.0233 \text{ or } .00507$$

X is calculated using both possible values of B.

   Solution 1: $B = 0.0233$

$$Z' = \frac{1}{0.0233 + 0.00592 - j0.0142} = 50.00 - j77.14$$

$$X = 77.14$$

   Solution 2: $B = 0.00507$

$$Z' = \frac{1}{0.00507 + 0.00592 - j0.0142} = 50.00 + j77.14$$

$$X = -77.14$$

Next, let us try to design the circuit shown in Figure 5-1(b). We have to test Y_oR_L to ensure it is less than 1.

$$Y_oR_L = \frac{25}{50} = 0.5$$

We can design a matching circuit with the topology shown in Figure 5-1(b).

$$X = -60 \pm \sqrt{(25)(50) - 25^2}$$

$$X = -60 \pm 25$$

We calculate Y' for both values of X.

Solution 3: $X = -35$

$$Y' = \frac{1}{25 + j60 - j35} = 0.0200 - j0.0200$$

$$B = 0.0200$$

Solution 4: $X = -85$

$$Y' = \frac{1}{25 + j60 - j85} = 0.0200 + j0.0200$$

$$B = -0.0200$$

We have only shown solutions for matching an impedance of admittance to a real impedance Z_o or admittance Y_o. It is possible to match to any arbitrary source with a slight change in the way the equations are derived. The real part of Y' or Z' must equal the real part of the source admittance or impedance. Then we can match the imaginary part by adding or subtracting some reactance with the final element. Using the circuit in Figure 5.1(a) to match a load to any source impedance ($R_s + jX_s$), we repeat the derivation by setting Re{Z'} = R_s. Equations 5.7 and 5.8 become

$$B = -B_L \pm \sqrt{\frac{G_L - R_s G_L^2}{R_s}} \qquad 5.16$$

$$Z = -\text{Im}\{Z'\} - X_s \qquad\qquad 5.17$$

Similarly, when using a circuit as shown in Figure 5.1(b) to match to a source admittance $(G_s + jB_s)$, Equations 5.13 and 5.14 become

$$X = X_L \pm \sqrt{\frac{R_L - G_s R_L{}^2}{G_s}} \qquad\qquad 5.18$$

$$Y = -\text{Im}\{Y'\} - B_s \qquad\qquad 5.19$$

These two circuit configurations are simple and effective methods of matching load impedances. Lumped-element matching is preferred for monolithic microwave integrated circuits because these circuits are smaller than those made with transmission line matching. When using discrete components, real capacitors and inductors usually do not function as ideal components at some frequencies. one has to be careful to ensure that any high-frequency effects are accounted for in the design. Most manufacturers of capacitors will provide the series resistance frequency of their capacitors. Some publish informative tutorials on the parasitic elements that are present at high frequencies.

5.3 Impedance Transformers

A single-series transmission line is a simple method of matching two impedances together, as shown in Figure 5-3. At microwave frequencies, the wavelength is in the order of a few centimeters or less. Conventional components and their interconnections become large compared to a wavelength. These parts do not function as lumped elements but as distributed components. Frequently, this has undesirable effects on the performance of the microwave circuit, i.e., the components become lossy and cease to function as pure inductors and capacitors. Therefore it is more efficient to use small sections of transmission lines instead of discrete circuit elements. The wavelengths of the signals in the circuit are often small compared to the overall size of the circuit housing. This section shows how to design an impedance-matching circuit using a single-series transmission line.

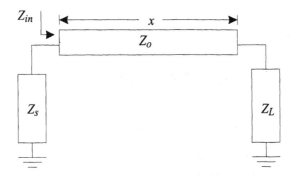

Figure 5-3 *A simple transmission line element for impedance matching.*

We will begin by presenting the general solution for matching two imped-ances with a single-series transmission line (see Figure 5-3). The impedance looking into a lossless transmission line was given in Equation 3.33,

$$Z_{in} = Z_t \frac{Z_L + jZ_t \tan \beta\ x}{Z_t + jZ_L \tan \beta\ x} \qquad 5.20$$

where Z_t is the impedance of the transmission line, x is the length, and $\beta = 2\pi/\lambda$. We assume the characteristic impedance of the transmission line is real

$$Z_{in}(Z_t + jZ_L \tan \beta x) = Z_t(Z_L + jZ_t \tan \beta x) \qquad 5.21$$

and separate the Equation 5.21 into the real part

$$R_s Z_t - R_s X_L \tan \beta x - R_L X_s \tan \beta x = Z_t R_L \qquad 5.22$$

and the imaginary part

$$R_s R_L \tan \beta x + X_s Z_t - X_s X_L \tan \beta x = Z_t X_L + Z_t^2 \tan \beta x \qquad 5.23$$

where $Z_{in} = R_s + jX_s$. There are two equations and two unknowns, x and Z_t. Solving first for $\tan \beta x$ then for Z_o, we find

$$\tan \beta x = \frac{Z_t(R_s - R_L)}{R_s X_L + R_L X_s} \qquad 5.24$$

We substitute Equation 5.24 into Equation 5.23 and solve for Z_t.

$$Z_t^2 \frac{(R_s - R_L)}{R_s X_L + R_L X_s} + (X_L - X_s) + (X_s X_L - R_s R_L) \frac{(R_s - R_L)}{R_s X_L + R_L X_s} = 0$$

$$Z_t = \sqrt{\frac{X_s^2 R_L - X_L^2 R_s}{R_s - R_L} + R_s R_L} \qquad \qquad 5.25$$

Z_t must be real; thus the quantity inside the square root must be positive.

Example 5.2: Find a transmission line that will match a $43 + j24$ ohm impedance with 75 ohms. We can find the impedance of the transmission line by the following:

$$Z_t = \sqrt{(43)(75) - \frac{75(24)^2}{(75\text{-}43)}} = \sqrt{1875}$$

$$= 43.3 \ \Omega$$

The length is found by solving for $\tan \beta x$.

$$\tan \beta x = \frac{\sqrt{1875}(75 - 43)}{(75)(24)}$$

$$= .7698$$
$$\beta x = .6561$$
$$x = .1044 \text{ wavelengths}$$

A common transmission line matching element is the *quarter-wave transformer*, which is frequently used when matching from one real impedance to another. When x is one-fourth of a wavelength, Equation 5.22 becomes

$$Z_s = Z_t \frac{Z_L + j Z_t \tan \dfrac{2\pi}{\lambda} \dfrac{\lambda}{4}}{Z_t + j Z_L \tan \dfrac{2\pi}{\lambda} \dfrac{\lambda}{4}} \qquad \qquad 5.26$$

$$Z_s = \frac{Z_t^2}{Z_L} \qquad \qquad 5.27$$

The characteristic impedance of the quarter-wave transmission line that matches Z_L to Z_s is

$$Z_t = \sqrt{Z_s Z_L} \qquad \qquad 5.28$$

The result can be seen on a Smith chart in Figure 5-4. A load impedance is plotted on the real axis at the point Z_L. Graphically, we find a point halfway between Z_L and the center of the Smith chart, Z_o. The impedance at this point is the impedance of the quarter-wave transmission line. The transmission line moves the point Z_L in a half circle, as shown in Figure 5-4, and comes to rest at Z_o.

Naturally, quarter-wave transformers are frequency-dependent and are designed to be one-fourth of a wavelength in length at a specific frequency. As the frequency of operation changes, the transmission line becomes a poorer approximation of one-fourth of a wavelength. The speed at which the match falls apart is related to the ratio of the load and source impedances that are being matched. We will show how to design broadband transformers using multiple cascaded transmission lines in Chapter 7.

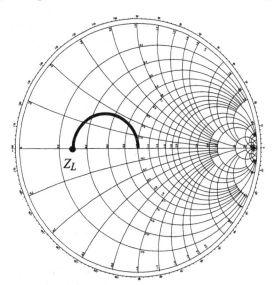

Figure 5-4 *A quarter-wave transformation plotted on a Smith chart. The load is matched to 50 ohms by traveling along the transmission line, as illustrated by the line.*

Example 5.3: Two 50 ohm loads are to be driven by a single 50 ohm source. We want equal power to be delivered to each load. Design a quarter-wave transformer so that when the two transmission lines are used, as shown in Figure 5-5, they will match both the loads to the source.

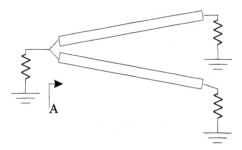

Figure 5-5 *Two quarter-wave transformers used to match two 50 ohm loads to one 50 ohm source.*

We want the impedance looking into the two transmission lines at point A to be 50 ohms. The combined impedance looking toward the two loads at point A result from two parallel impedances. The two arms of the power splitter must be identical to allow for an equal power split between the two 50 ohm loads. The transmission lines must be designed so that the following two conditions are met: the impedance Z_{in} looking into each arm of the power splitter at point A must be the same and the parallel combination of the impedances of the two arms must equal 50 ohms. Thus, the transmission lines must transform the 50 ohm load impedances so that Z_{in} = 100 ohms. The characteristic impedance of the quarter-wave transformers has to be

$$Z_t = \sqrt{(50)(100)} = 70.7\Omega$$

5.4 Single-Shunt Stub Matching Circuits

This section shows how to design impedance-matching circuits using shunt transmission lines, or stubs. We will be able to construct simple impedance-matching circuits using a single-shunt transmission line that is correctly positioned somewhere along a transmission line leading to a load, as shown in Figure 5-6. If we are working in a 50 ohm system, an open or shorted shunt transmission line, or stub, is placed with a specific length of 50 ohm line between the stub and the load impedance. We can match any arbitrary impedance with this kind of circuit topology. Furthermore, the impedance of the transmission lines can equal the characteristic impedance of the system.

The decision whether to use an open or shorted stub will depend primarily on the type of transmission line that is used. An open stub is preferred when using microstrip lines because shorted lines are more difficult to fabricate. Shorted stubs are preferred when using coaxial cable because the coaxial short circuits are more perfect than the open circuits.

First, let us look at the admittance of an open and shorted transmission line. The admittance of a shunt open stub is given in Equation 5.20 where Z_L is infinite.

$$Y = \frac{1}{Z} = \frac{1}{Z_o} j \tan \frac{2\pi x}{\lambda} \qquad 5.29$$

The admittance looking into a shorted stub is given in Equation 5.20 where Z_L is equal to zero.

$$Y = \frac{1}{Z} = -\frac{1}{Z_o} j \cot \frac{2\pi x}{\lambda} \qquad 5.30$$

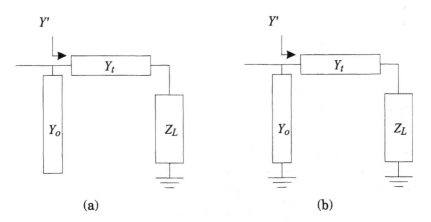

(a)   (b)

Figure 5-6 *Impedance-matching circuits using a single open stub (a) or a shorted stub (b).*

We can better understand these transmission line components by looking at a Smith chart. An open stub can be shown by beginning at the right hand edge at the open end of the chart and traveling clockwise around the edge of the Smith chart that corresponds to the length of the line, as shown in Figure 5-7(a). A shorted stub is plotted by starting at the left-hand edge at the open end of the Smith chart and traveling clockwise around the edge [Figure 5-7(b)].

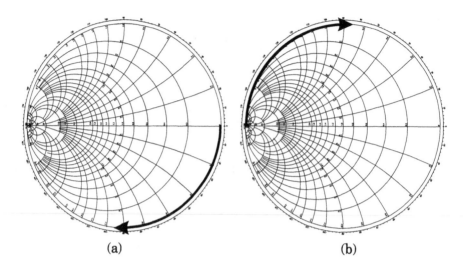

(a) (b)

Figure 5-7 *An open (a) and shorted (b) transmission line plotted on the admittance Smith chart.*

Open or shorted transmission lines can only produce an imaginary imped-ance. Since the transmission line is connected in shunt, it can only add some imaginary admittance to the admittance looking in at the location of the stub. The key to designing the single-stub matching circuit is to make the real part of the admittance looking into the transmission line at the stub equal to Y_o. Any imaginary part is then canceled with the open or shorted stub.

Figure 5-8 shows a matching circuit using an open stub as the tuning element. Suppose we want to match the load to a transmission line with a characteristic impedance of $Z_o = 1/Y_o$. The series transmission line is designed to make the real part of Y' equal to Y_o. Looking into this trans-mission line before the stub, the admittance should be $Y_o + jB$ where B can be any value. The length of the stub is designed so that the shunt admit-tance cancels the imaginary part of Y'.

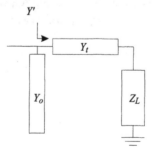

Figure 5-8 *A single-stub matching circuit using an open stub.*

On a Smith chart, the series transmission line rotates the admittance of the load until it lands on the constant Y_o conductance circle, as shown in Figure 5-9. When the admittance looking into the series transmission line is on this circle, the shunt stub can be designed to cancel the remaining imaginary part.

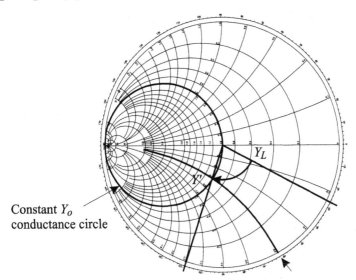

Constant Y_0
conductance circle

Figure 5-9 *A series transmission line is used to move the load admittance to the constant Y_o circle on the Smith chart to Y'. Eventually, the imaginary part can be removed by the stub.*

Looking into the series transmission line at Y', $Z_L = R_L + X_L$ has been transformed to

$$Y' = \frac{1}{Z'} = \frac{1}{Z_t} \frac{Z_t + jZ_L \tan \beta x}{Z_L + jZ_t \tan \beta x} \qquad 5.31$$

where Z_t is the characteristic impedance of the transmission line and x is the length of the transmission line. For convenience, the characteristic impedance of the transmission line is equalized to Y_o. The length of this line is designed so that the real part of Equation 5.31 is equal to Y_o The admittance at Y' looking into the series transmission line is

$$Y' = Y_o \frac{\left[(Z_0 - X_L \tan \beta x) + jR_L \tan \beta x\right]\left[R_L - j(X_L + Z_o \tan \beta x)\right]}{R_L^2 + (X_L + Z_o \tan \beta x)^2} \qquad 5.32$$

We then take the real part, force it to equal Y_o and solve for $\tan \beta x$

$$Y_o = \mathrm{Re}\{Y'\} = Y_o \frac{R_L \left(1 + \tan^2 \beta x\right)}{R_L{}^2 + \left(X_L + Z_o \tan \beta x\right)^2} \qquad 5.33$$

$$Z_o \left(R_L - Z_o\right) \tan^2 \beta x - 2X_L Z_o \tan \beta x + \left(R_L Z_o - R_L^2 - X_L^2\right) = 0 \qquad 5.34$$

$$\tan \beta x = \frac{X_L \pm \sqrt{Y_o R_L \left[\left(Z_o - R_L\right)^2 + X_L^2\right]}}{R_L - Z_o} \qquad 5.35$$

except when $R_L = Z_o$. When $R_L = Z_o$, Equation 5.35 is evaluated as $(R_L - Z_o)$ approaches zero and becomes

$$\tan \beta x = \frac{-X_L}{2Z_o} \qquad 5.36$$

Equation 5.35 shows two solutions for $\tan \beta x$ corresponding to both sides of the constant Y_o circle. Once the length of the series transmission line is known, the imaginary part of Equation 5.27 can be found.

$$\mathrm{Im}\{Y'\} = \frac{R_L^2 \tan \beta x - \left(Z_o - X_L \tan \beta x\right)\left(X_L + Z_o \tan \beta x\right)}{Z_o \left[R_L{}^2 + \left(X_L + Z_o \tan \beta x\right)^2\right]} \qquad 5.37$$

The length of the shunt stub can be found using Equation 5.29 or 5.30 by making $Y = \mathrm{Im}\{Y'\}$.

Example 5.4: Design two single-stub matching circuits to match a $36 + j48$ ohm load to a 50 ohm source using an open stub for the first one and a shorted stub for the second one. The reflection coefficient is

$$\Gamma = \frac{36 + j48 - 50}{36 + j48 + 50} = -0.1134 + j0.4948 = 0.5077 @ 77.09°$$

This load admittance is plotted on the Smith chart in Figure 5-10. A 50 ohm transmission line is needed to rotate the point to the .02 mho conductance circle. Using Equation 5.35, we calculate $\tan \beta x$

$$\tan \beta x = \frac{48 \pm \sqrt{.02 * 36\left[(50 - 36)^2 + 48^2\right]}}{36 - 50}$$

$$= -3.4286 \pm 3.0305$$

in order for tan βx to take one of two values, -0.3981 or -6.4590. When tan βx is less than zero, the length of the transmission line will be negative. In this case, we add half a wavelength to obtain a positive length.

$$x = 0.4397 \quad \lambda = 158.3° \qquad \text{or} \qquad x = 0.2744 \quad \lambda = 98.8°$$

A transmission line that is 0.4397 wavelengths long will move Y_L to the .02 mho conductance circle, as shown in Figure 5-10. The admittance at that point is

$$\text{Im}\{Y'\} = \frac{36^2 * (-0.3981) - (50 - 48 * (-0.3981))(48 + 50 * (-0.3981))}{50 * \left[36^2 + (48 + 50 * (-0.3981))^2 \right]}$$

The admittance of the open stub should be

$$B = -\text{Im}[Y'] = 0.02357$$

An open stub is preferred in this case because it will be shorter than a shorted stub and will make the matching circuit somewhat less sensitive to variations in frequency (Figure 5-11).

$$0.0236 = 0.02 \tan \frac{2\pi x}{\lambda}$$

$$x = 0.1381 \quad \lambda = 49.7°$$

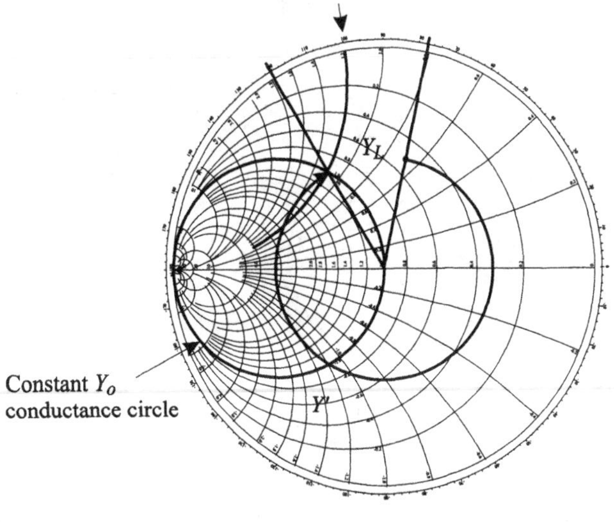

Figure 5-10 *Designing the match circuit in Example 5.4 on the Smith chart.*

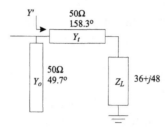

Figure 5-11 *The matching circuit in Example 5.4 using an open stub.*

A similar circuit with an open stub can be designed by utilizing a series transmission line that is 0.2744 wavelengths long. The length of the series transmission line has been calculated so that the admittance at Y' is on the lower side of the 0.02 mho constant conductance circle, as shown in Figure 5-12. The imaginary part of Y' is

$$\text{Im}\{Y'\} = \frac{36^2 * (-6.4590) - (50 - 48 * (-6.4590))(48 + 50 * (-6.4590))}{50 * \left[36^2 + (48 + 50 * (-6.4590))^2\right]}$$

$$= 0.2357$$

Then a short-circuit stub is used to subtract the imaginary part.

$$0.0236 = 0.02 \tan \frac{2\pi x}{\lambda}$$

$$x = -0.1120 \qquad \lambda = 40.4°$$

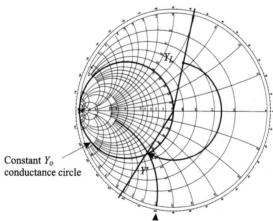

Figure 5-12 *The load admittance is rotated 98.8° and comes to a conclusion on the 0.02 mho conductance circle.*

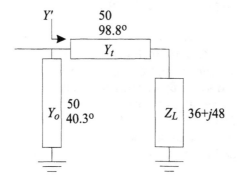

Figure 5-13 *The matching circuit of Example 5.4 using a shorted stub.*

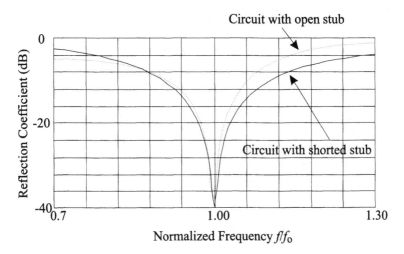

Figure 5-14 *The reflection coefficient of the two matching circuits as a function of frequency as modeled by Eagleware's =SuperStar=.*

Single-stub matching circuits are a simple and effective method using only Z_o impedance transmission lines. The length of a series transmission line is designed to move the load admittance to the constant Y_o conductance circle on the Smith chart. Then a shunt stub is designed to remove the imaginary part. This type of matching circuit is graphically and mathematically easy to design.

5.5 Single-Series Stub Tuning

Series stubs can be convenient to use when working with coaxial cable transmission lines. Designing impedance-matching circuits using series open or shorted transmission lines is similar to the procedure shown in the previous section. Instead of working with admittance, we will be working with impedance. The series transmission line is designed to transform the load impedance so that it has a real part equal to the characteristic impedance Z_o. These two matching topologies are shown in Figure 5-15. We will use the impedance Smith chart when plotting the load impedance. Then we will rotate the load impedance around the Smith chart until it intersects the constant Z_o circle. The imaginary part is canceled out by adding a series short-circuited stub with a length x, thus satisfying the equation

$$\text{Im}\{Z'\} = -Z_o \tan \beta x \qquad\qquad 5.38$$

or using an open-circuited stub with a length that satisfies

$$\text{Im}\{Z'\} = -Z_o \cot \beta x \qquad\qquad 5.39$$

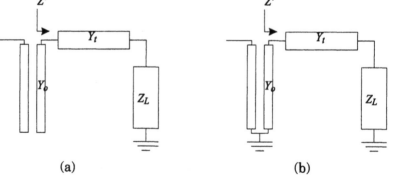

(a)   (b)

Figure 5-15 *Two impedance-matching circuit topologies using a shorted series stub (a) and an open shorted stub (b).*

The process is illustrated on the impedance Smith chart in Figure 5-16. The impedance at Z' is given by Equation 5.20 with $Z_L = R_L + X_L$. We solve for the impedance Z' looking into the series transmission line.

$$Z' = \frac{1}{Y_0} \frac{Y_o + jY_L \tan \beta x}{Y_L + jY_o \tan \beta x} \qquad\qquad 5.40$$

We find the real part of Equation 5.38 and adjust it to equal Z_o.

$$Z_o = \text{Re}\{Z'\} = Z_o \frac{G_L(1+\tan^2 \beta x)}{G_L^2(B_L + Y_o \tan \beta x)^2} \qquad 5.41$$

$$Y_o(G_L - Y_o)\tan^2 \beta x - 2B_L Y_o \tan \beta x + \left(G_L Y_o - G_L^2 - B_L^2\right) = 0 \qquad 5.42$$

$$\tan \beta x = \frac{B_L \pm \sqrt{Z_o G_L\left[(Y_o - G_L)^2 + B_L^2\right]}}{G_L - Y_o} \qquad \begin{array}{l} G_L \neq Y_o \\ \\ \end{array} \qquad 5.43$$

$$\tan \beta x = \frac{-B_L}{2Y_o} \qquad (G_L = Y_o)$$

The length of the series transmission line is found by solving Equation 5.39. The length is used to find the imaginary part.

$$\text{Im}\{Z'\} = \frac{G_L^2 \tan \beta x - (Y_o - B_L \tan \beta x)(B_L + Y_o \tan \beta x)}{Y_o\left[G_L{}^2 + (B_L + Y_o \tan \beta x)^2\right]} \qquad 5.44$$

Next, the length of the series open stub can be found by

$$-\text{Im}\{Z'\} = Z_o \tan \beta x \qquad 5.45$$

The impedance of a series open stub is

$$-\text{Im}\{Z'\} = Z_o \cot \beta x \qquad 5.46$$

Example 5.5: Design a circuit to match a $35 - j15$ ohm load to a 50 ohm transmission line using a series stub and open stub topology.

$$Y = \frac{1}{35 - j15} = 0.02414 + j0.01034$$

The reflection coefficient is

$$\Gamma = -0.1409 - j2013 = 0.2458@125.0°$$

This load impedance is plotted on the Smith chart in Figure 5-17. A 50 ohm transmission line is needed to rotate the point to the 50 ohm resistance circle. Equation 5.41 is used to find tan βx.

$$\tan \beta x = \frac{(0.01034) \pm \sqrt{(50)(0.02414)\left[(0.02-0.02414)^2 + 0.01034^2\right]}}{(0.02414)-(0.02)}$$

$$= 2.500 \pm 2.958$$
$$x = .2212\lambda \text{ or } 0.4316\lambda$$

The series transmission line can be 79.62° or 155.39° long. A transmission line that is 155.4 electrical degrees in length will move Z_L to the 50 ohm resistance circle. The resistance at that point is found using Equation 3.35.

$$Z = 50\frac{35-j15+j50*(5.458)}{50+j(35-j15)*(5.458)} = 50 + j25.355$$

A shorted series stub is needed to remove the imaginary part. The length of a shorted stub is

$$25.255 = 50 \tan \beta x$$
$$x = 0.0747 \quad \lambda = 27.9°$$

The matching circuit is shown in Figure 5-18.

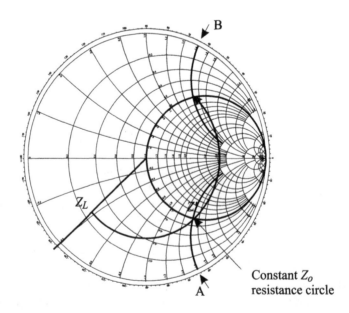

Figure 5-16 *Matching a load impedance Z_L with a series stub transmission line circuit.*

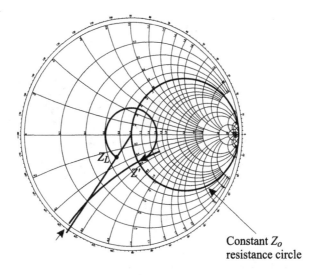

Figure 5-17 *Designing the match circuit in Example 5.5 on the Smith chart.*

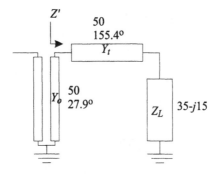

Figure 5-18 *The matching circuit in Example 5.5 using a series shorted stub.*

A similar circuit can be designed using an open series stub by rotating the load impedance to the top of the constant Z_o resistance circle. The length of the series transmission line is chosen so that the impedance is on the upper side of the constant conductance circle, as shown in Figure 5-19. The transmission line should be 79.62 electrical degrees in length for the impedance looking in at Z' to be $50\Omega - j1.004$. An open-circuit stub is used to subtract the imaginary part.

$$-25.3546 = 50 \tan \beta x$$
$$x = 0.1753 \quad \lambda = 63.11°$$

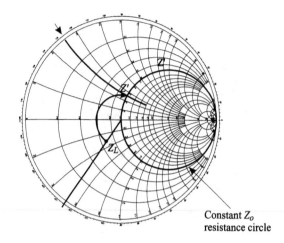

Constant Z_o
resistance circle

Figure 5-19 *Using an open series stub (Example 5.5), the load impedance is rotated and appears on the upper side of the 50 ohm resistance circle.*

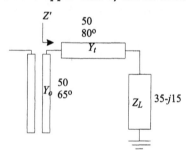

Figure 5-20 *The matching circuit of Example 5.5 using a series open stub.*

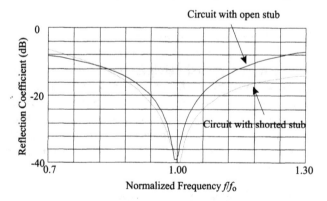

Figure 5-21 *A plot of the reflection coefficient of both circuits in Example 5.5 as a function of frequency.*

5.6 Double-Stub Matching

Some impedances are difficult to match using single-stub tuners, series transmission lines, or lumped elements. We will find that, under some conditions, a theoretical circuit design is difficult or even impossible to implement in physical form. The transmission line stub may be too short or too long, or an impedance transformer may be difficult to fabricate. For example, there may be times when we want to match a load with a very high reflection coefficient. Let us experiment and use a single-stub tuner. We add a series transmission line to rotate the load admittance so that it intersects the constant conductance Y_o circle as usual. The load admittance will arrive on the Y_o circle near the shorted end of the Smith chart. The susceptance at this point is very high and the shunt transmission line would be almost one-quarter wavelength long. The quality of the match will be very sensitive to a small variation in the length of the shunt element. Furthermore, any junction effects between the series and shunt transmission lines could also have a serious effect on the match.

One solution for this problem is to use a double-stub topology, as shown in Figure 5-22. The shunt transmission line stub near the load cancels some of the imaginary part at Y' before the series transmission rotates the admittance to the Y_o circle. The admittance Y'' can be brought closer to the center of the Smith chart than it would normally be without the stub B_1. The residual susceptance at Y'' may not be as difficult to remove. We can think of the double-stub circuit as a single-stub circuit with a little help from an extra stub at the load.

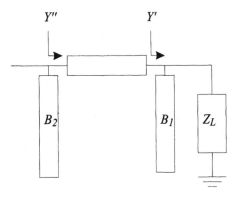

Figure 5-22 *The topology of a double-stub matching circuit.*

The double-stub matching network uses two open or short-circuited shunt transmission lines separated by a series transmission line. The goal is to arrive at the constant Y_o circle by the end of the series transmission line. The admittance at the first stub Y' should lie on a constant Y_o circle. However, this constant Y_o circle is not in its usual location on the Smith chart. Let us assume that the series transmission line is 3/8 of a wavelength long. This causes the constant Y_o circle to be rotated to the right side of the Smith chart, as shown in Figure 5-23. The first shunt stub removes some of the imaginary component of the load admittance. This results in an admittance Y' that is closer to the center of the Smith chart making it easier to match. The quarter-wavelength series transmission line moves the admittance from Y' to Y''. The second shunt stub cancels any remaining imaginary part of Y''.

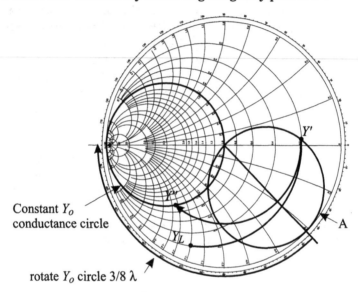

Figure 5-23 *The constant Y_o conductance circle is rotated by the second series transmission line.*

There are many degrees of freedom in the synthesis of this kind of matching circuit. First, there are the impedances and lengths of the three transmission lines. Usually, the impedances of all the transmission lines are fixed at some convenient impedance such as Z_0. Second, the length of the series transmission line is chosen by inspecting the location of the load on the Smith chart. It takes a little practice to determine a convenient location for the constant Y_o circle before it is rotated by the series transmission line. The length of the series transmission line is easy to choose after a little experimentation.

The analytical solution for the double-stub matching circuit is very similar to the one in Section 5.3 for the single-shunt stub circuit. The admittance Y' is the sum of the load admittance $Y_L = 1/Z_L$ and the transmission line stub admittance.

$$Y' = Y_L + jB_1 \qquad 5.47$$

The admittance looking toward the load at Y'' is the admittance in Equation 5.47 transformed by the series transmission line

$$Y'' = Y_t \frac{(Y_L + jB_1) + jY_t \tan \beta x}{Y_t + j(Y_L + jB_1) \tan \beta x} \qquad 5.48$$

where Y_t is the admittance of the transmission line, and x is the length. The real part of this admittance should be equal to the real part of the source admittance.

$$\text{Re}\{Y_s\} = \text{Re}\{Y''\} = \text{Re}\left\{ Y_t \frac{(Y_L + jB_1) + jY_t \tan \beta x}{Y_t + j(Y_L + jB_1) \tan \beta x} \right\} \qquad 5.49$$

$$= \text{Re}\left\{ \frac{Y_t[(G_L + j(B_L + B_1 + Y_t \tan \beta x)][Y_t - (B_1 + B_L) \tan \beta x - jG_L \tan \beta x)]}{[Y_t - (B_1 + B_L) \tan \beta x]^2 + (G_L \tan \beta x)^2} \right\} \qquad 5.50$$

$$= \frac{Y_t\{G_L[Y_t - (B_1 + B_L) \tan \beta x] + (B_L + B_1 + Y_t \tan \beta x)(G_L \tan \beta x)]}{[Y_t - (B_1 + B_L) \tan \beta x]^2 + (G_L \tan \beta x)^2} \qquad 5.51$$

We force Equation 5.49 to equal Y_o, which yields a quadratic equation for B_1.

$$B_1^2 \tan^2 \beta x - 2B_1(Y_t - B_L \tan \beta x) \tan \beta x + $$

$$(Y_t - B_L \tan \beta x)^2 - G_L Y_t(1 - \tan^2 \beta x) = 0 \qquad 5.52$$

$$B_1 = \frac{(Y_t - B_L \tan \beta x) \pm \tan \beta x \sqrt{Y_t G_L - G_L^2}}{\tan \beta x} \qquad 5.53$$

There are three degrees of freedom in Equation 5.53, Y_t, x, and B_1 and only one independent variable, $\text{Re}\{Y_s\}$. Commonly, the characteristic impedance and length of the series transmission line is selected by experimenting with the rotated Y_o circle on the Smith chart. After calculating B_1 with Equation 5.53, B_2 is found by taking the negative of the imaginary part of Y''.

$$B_2 = -\text{Im}\left\{ \frac{Y_t[(G_L + j(B_L + B_1 + Y_t \tan \beta x)][Y_t - (B_1 + B_L) \tan \beta x - jG_L \tan \beta x)]}{[Y_t - (B_1 + B_L) \tan \beta x]^2 + (G_L \tan \beta x)^2} \right\} \qquad 5.54$$

$$B_2 = \frac{Z_t \left(G_L^2 + (B_L + B_1)^2\right) \tan \beta x + (B_L + B_1 - Y_t)\tan^2 \beta x - (B_L + B_1)}{\left(Y_t - (B_L + B_1)\tan \beta x\right)^2 + (G_L \tan \beta x)^2} \qquad 5.55$$

Example 5.6: Use a double-stub circuit to match the load in Example 5.4, $36 + j48$ Ω with only 50 ohm transmission lines and open-circuit transmission lines. This example shows how to design a smaller circuit by using a double-stub matching circuit rather than a single-stub circuit. Figure 5-24 shows the load plotted on the admittance Smith chart. The admittance of the load is

$$Y_L = .0100 - j0.0133$$

If we use a series transmission line one-quarter of a wavelength long, it is easy to move the load to the pre-rotated Y_o circle shown as a thick circle on the right side of the Smith chart in Figure 5-24. Admittances on this circle will rotate around the Smith chart by the series transmission line and will end up on the constant Y_o conductance circle. When $x = 1/4$ λ, tan βx is infinite. Equation 5.53 is evaluated when it approaches infinity.

$$B_1 = -B_L \pm \sqrt{G_L Y_t - G_L^2}$$

$$B_1 = 0.0133 \pm 0.0100$$

$$B_1 = 0.0233 \text{ or } 0.00333$$

Figure 5-24 shows that the first shunt stub moves the load admittance to the bottom part of the rotated Y_o circle. We need to use the larger of the two solutions for B_1. This corresponds to a stub length of

$$x = \frac{\tan^{-1}(Z_o * 0.0233)}{2\pi}$$

$$x = 0.1372 \qquad \lambda = 49.4°$$

The series transmission line rotates the admittance from Y' to Y''.

$$Y'' = \frac{0.02^2 (0.010 - j(-0.0133 + .0233))}{0.010^2 + (-0.0133 + 0.0233)^2}$$

$$= 0.02 - j0.02$$

A second shunt stub with a susceptance of 0.326 mhos is needed to bring
the match to .02 mhos.

$$x = \frac{\tan^{-1}(Z_o * 0.0200)}{2\pi}$$

$$x = 0.125 \qquad \lambda = 45°$$

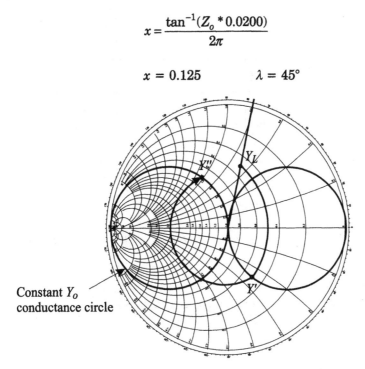

Figure 5-24 *The admittance Smith chart for double-stub matching in
Example 5.6.*

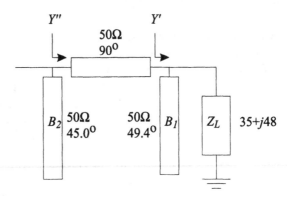

Figure 5-25 *The matching circuit of Example 5.6, using a double-stub
matching circuit.*

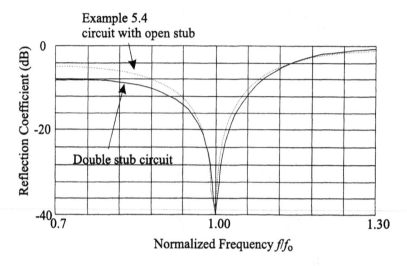

Figure 5-26 *A plot of the frequency response of the matching circuit in Example 5.6.*

The frequency response of this circuit is plotted with the response of the open-stub matching circuit in Example 5.4. The double-stub circuit offers a small improvement in bandwidth.

5.7 Summary

In Chapters 1, 2 and 3, we have described circuits with network parameters and analyzed existing circuits. In this chapter, we have introduced some useful synthesis techniques and shown how to design a few simple and effective impedance-matching circuits. We have not addressed the synthesis problem in general since many helpful books on the subject exist (see [1, 2]). Impedance-matching circuits are needed to match a transistor to a transmission line when designing an amplifier. They are used at the input and output of a transistor so that inserting it into a transmission line system yields a desired result, i.e., maximum gain. For example, we might like a circuit to amplify a signal without causing reflections in the transmission system. Matching circuits are used to transform the input and output impedance of the transistor to the characteristic impedance of the transmission line. In this chapter, we discussed four circuit types used to match one impedance to another.

Section 5.2 showed how a simple matching network consisting of two lumped elements can match any arbitrary impedance to another. These circuits are needed for narrowband matching when discrete capacitors and inductors are used. They are an effective tool when designing amplifiers at RF frequencies and on monolithic integrated circuits.

Section 5.3 showed a matching technique using a single-series transmission line. Microwave circuits constructed from microstrip or another transmission medium primarily use transmission lines and stubs to match impedances. A series transmission line is the simplest of these distributed matching networks. With the proper choice of impedance and electrical length, any two impedances can be matched. The quarter-wave transmission line, which matches any two real impedances, constitutes a special case. However, sometimes it is not possible to manufacture the physical circuit and other circuit topologies have to be used.

Section 5.4 explained how to use a series line and a shunt stub as a matching circuit. In this circuit, we employ transmission lines of a predetermined impedance and calculate their length. Similarly, in Section 5.5 we used the same premises except that, instead of a shunt stub, a series open-circuit or short-circuit stub is used. These circuit topologies broaden our choices of matching circuits. The derivations given here show how to match a load to a real impedance, which is usually the characteristic impedance of the transmission line system. With only a small change in the derivation, any source impedance can be matched. This is useful when matching the output of one transistor to the input of another in a two-stage amplifier.

Finally, Section 5.6 introduced double-stub matching circuits and presented one of the possible circuit types that use three transmission lines. The other types of three-element matching circuits can be designed using the instructions for the double-stub circuits as a guide.

The networks and techniques in this chapter apply to narrowband circuits. The synthesis techniques assume a single frequency of operation. The lumped-element capacitors and inductors are found by their reactance or susceptance. This implies single frequency excitation. Likewise, we synthesize transmission line matching elements that have a certain electrical length that is frequency-dependent. Broadband design methods are discussed in Chapter 7.

5.8 Problems

5.1 Design a matching circuit using the topology in Figure 5-1(a) to match a $22 + j100$ ohm load to 50 ohms.

5.2 Design a matching circuit using the topology in Figure 5-1(a) to match a $136 - j40$ ohm load to a $75 + j34$ ohm source.

5.3   Design a matching circuit using the topology in Figure 5-1(b) to match a $25 + j15$ ohm load to 100 ohms.

5.4   Using the series transmission line circuit shown in Figure 5-3, find the impedance and length of a transmission line to match a $36 + j110$ ohm load to 75 ohms.

5.5   Design a power splitter, such as the one shown in Figure 5-5, which equally splits power three ways instead of two.

5.6   Show how two series transmission lines, one of which is a quarter-wave transformer, can be used to match a load impedance to 50 ohms.

5.7   Using the single-shunt stub transmission line circuit shown in Figure 5-6(a), design a circuit to match a $25 - j35$ ohm load to 75 ohms.

5.8   Find new design Equations 5.33 and 5.35 for the single-shunt stub circuit in Figure 5-6(a) so that any arbitrary source impedance can be matched to any load.

5.9   Using the single-shunt stub transmission line circuit shown in Figure 5-6(b), design a circuit to match a $115 + j32$ load to 75 ohms.

5.10 Find new design Equations 5.41 and 5.42 for the single-series stub circuit in Figure 5.6(b) so that any arbitrary source impedance. can be matched.

5.11 Design a double-stub tuner to match a $400 - j16$ ohm load to 50 ohms using a quarter wavelength 50 ohm series transmission line.

5.12 Design a double-stub matching circuit with a series transmission line that is $3/8\lambda$ long and matches a $32 - j12$ ohm load to a $95 + j30$ ohm source impedance.

5.9 References

1.   Collin, R. E. *Foundations for Microwave Engineering*. New York: McGraw-Hill, 1966.

2.   Mathaei, George L., Leo Young, and E. M. T. Jones. *Microwave Filters, Impedance-Matching Networks, and Coupling Structures*. New York: McGraw-Hill, 1964.

5.10   Appendix

MatLab programs for matching circuit design.

Lumped-element matching circuits:

lumped.m

```
function[b,x]=lumped(zl,zs)
% This routine finds the lumped matching circuit
% elements given a load impedance zl and a source
% impedance zs which should be real
% Usage [b(4),x(4)]=lumped(zl,zs) where
% zl is the load impedance
% zs is the source impedance
j=sqrt(1);
rl=real(zl);
xl=imag(zl);
yl=1/zl;
gl=real (yl);
bl=imag (yl);
disp('Solution one (shunt reactance nearest the load):')
if (zs*gl < 1.)
  b(1)=-1*bl-sqrt(gl/zs-gl^2);
  b(2)=-1*bl+sqrt(gl/zs-gl^2);
  x(1)=imag (1/(j*b(1)+yl));
  x(2)=imag(1/(j*b(2)+yl));
  sprintf ('The shunt reactance is %0.5g or %0.5g,\n',b(1),b(2))
    sprintf ('with a series reactance of %.5g or %0.5g respectively
\n',x(1),x(2))
  else
  disp ('Is not possible with this load and source')
  b(1)=+Inf;
  b(2)=-Inf;
  x(1)=0;
  x(2)=0;
  end
  disp ('Solution two (series reactance nearest the load):')
  if (rl/zs < 1. )
  x(3)=-1*xl-sqrt(zs*rl-rl^2)
  x(4)=-1*xl+sqrt(zs*rl-rl^2)
  b(3)=imag(1/(zl+j*x(3)))
  b(4)=imag(1/(zl+j*x(4)))
```

```
    sprintf('The series reactance is %0.5g or %0.5g,\n',x(3),x(4))
      sprintf('with a shunt reactance of %.5g or %0.5g respectively
\n',b(3),b(4))
   else
   disp('Is not possible with this load an source')
   b(3)=+Inf;
   b(4)=-Inf;
   x(3)=0;
   x(4)=0;
   end
```

Transmission line transformer method:

transfm.m

```
function[zt,el]=transfm(zl,zs)
% This function finds the characteristic impedance
% and electrical length of an impedance transformer
% Usage [z,x]=transfm(zl,zs) where
% zl is the load impedance
% zs is the source impedance
rl=real(zl);
xl=imag(zl);
rs=real(zs);
xs=imag(zs);
zt2=rs*rl-(xs^2*rl+xl^2*rs)/(rs-rl);
if (zt2 < 0 )
   disp('No solution possible');
else
   zt=sqrt(zt2);
   beta=atan(zt*(rs-rl)/(rs*xl-rl*xs));
   x=beta/(2*pi);
   sprintf('Characteristic impedance is %0.5g.',zt)
   sprintf('with a length of %0.5g wavelengths.',x)
end
```

Single-shunt stub matching method:

shunt1.m

```
function[l1,l2]=shunt1(zl,zo)
% This function finds the lengths of a single-shunt
% stub matching circuit.
```

```
% Usage: [l1(2),l2(4)]=shunt1(zl,zo)where
%   zl is the load impedance
%   zo is the source impedance and the impedance
%     of the transmission lines.
% Function returns lengths in fractions of a wavelength.
%   l1(2) are the lengths of the two series solutions
%   l2(4) are the shunt stub lengths for each solution of l1
%       l2(1) is solution 1 open stub length,
%       l2(2) is solution 1 shorted stub length,
%       l2(3) is solution 2 open stub length,
%       l2(4) is solution 2 shorted stub length.
j=sqrt(-1);
rl=real(zl);
xl=imag(zl);
yl=1/zl;
gl=real(yl);
bl=imag(yl);
yo=1/zo;
if (abs(rl-zo) < 1e-5)
  disp('There is only one solution.')
  tan βx=rl/2*zo;
  l1(1)=atan(tan βx)/(2*pi);
  if ( l1(1) < 0 )
    l1(1)=l1(1) + 0.5;
  end
  sprintf('The series transmission line is %0.5g wavelengths long.',l1(1))
  yp=yo*(yl+j*yo*tan βx)/(yo+j*yl*tan βx);
  b=-1*imag(yp);
  l2(1)=atan(b*zo)/(2*pi);
  if ( l2(1) < 0 )
    l2(1)=l2(1)+0.5;
  end
  sprintf('Use an open stub %0.5g wavelengths long.',l2(1))
  l2(2)=-1*acot(zo*b)/(2*pi);
  if ( l2(2) < 0 )
    l2(2)=l2(2)+0.5;
  end
  sprintf('Use a shorted stub %0.5g wavelengths long.',l2(2))
else
  tan βxp=(xl+sqrt(yo*rl*((zo-rl)^2+xl^2)))/(rl-zo);
  l1(1)=atan(tan βxp)/(2*pi);
  if ( l1(1) < 0 )
    l1(1)=l1(1) + 0.5;
```

```
    end
    tan βxn=(xl-sqrt(yo*rl*((zo-rl)^2+xl^2)))/(rl-zo);
    l1(2)=atan(tan βxn)/(2*pi);
    if ( l1(2) < 0 )
       l1(2)=l1(2) + 0.5;
    end
    disp('There are two solutions.')
    sprintf('The series transmission line can be %.05g wavelengths
long.',l1(1))
    yp=yo*(yl+j*yo*tan βxp)/(yo+j*yl*tan βxp);
    b=-1*imag(yp);
    l2(1)=atan(b*zo)/(2*pi);
    if ( l2(1) < 0 )
       l2(1)=l2(1)+0.5;
    end
    sprintf('Use an open stub %0.5g wavelengths long.',l2(1))
    l2(2)=-1*acot(zo*b)/(2*pi);
    if ( l2(2) < 0 )
       l2(2)=l2(2)+0.5;
    end
    sprintf('Or, use a shorted stub %0.5g wavelengths long.',l2(2))
    sprintf('Solution two is a series transmission line %.05g wavelengths
long.',l1(2))
    yp=yo*(yl+j*yo*tan βxn)/(yo+j*yl*tan βxn);
    b=-1*imag(yp);
    l2(3)=atan(b*zo)/(2*pi);
    if ( l2(3) < 0 )
       l2(3)=l2(3)+0.5;
    end
    sprintf('Use an open stub %0.5g wavelengths long.',l2(3))
    l2(4)=-1*acot(zo*b)/(2*pi);
    if ( l2(4) < 0 )
       l2(4)=l2(4)+0.5;
    end
    sprintf('Or, use a shorted stub %0.5g wavelengths long.',l2(4))
    end
```

Single-series stub matching method:

series1.m

```
function[l1,l2]=series1(zl,zo)
% This function finds the lengths of a single-shunt
```

```
% stub matching circuit.
% Usage: [l1(2),l2(4)]=shunt1(zl,zo)where
%   zl is the load impedance
%   zo is the source impedance and the impedance
%     of the transmission lines.
% Function returns lengths in fractions of a wavelength.
%   l1(2) are the lengths of the two series solutions
%   l2(4) are the shunt stub lengths for each solution of l1
%     l2(1) is solution 1 open stub length,
%     l2(2) is solution 1 shorted stub length,
%     l2(3) is solution 2 open stub length,
%     l2(4) is solution 2 shorted stub length.
j=sqrt(-1);
rl=real(zl);
xl=imag(zl);
yl=1/zl;
gl=real(yl);
bl=imag(yl);
yo=1/zo;
if (abs(gl-yo) < 1e-5)
  disp('There is only one solution.')
  tan βx=gl/2*yo;
  l1(1)=atan(tan βx)/(2*pi);
  if ( l1(1) < 0 )
    l1(1)=l1(1) + 0.5;
  end
  sprintf('The series transmission line is %0.5g wavelengths long.',l1(1))
  zp=zo*(yo+j*yl*tan βx)/(yl+j*yo*tan βx);
  x=-1*imag(zp);
  l2(1)=atan(-1*x*yo)/(2*pi);
  if ( l2(1) < 0 )
    l2(1)=l2(1)+0.5;
  end
  sprintf('Use a shorted series line of %0.5g wavelengths long.',l2(1))
  l2(2)=acot(yo*x)/(2*pi);
  if ( l2(2) < 0 )
    l2(2)=l2(2)+0.5;
  end
  sprintf('Use an open series line of %0.5g wavelengths long.',l2(2))
else
  tan βxp=(bl+sqrt(zo*gl*((yo-gl)^2+bl^2)))/(gl-yo);
  l1(1)=atan(tan βxp)/(2*pi);
  if ( l1(1) < 0 )
```

```
      l1(1)=l1(1) + 0.5;
   end
   tan βxn=(bl-sqrt(zo*gl*((yo-gl)^2+bl^2)))/(gl-yo);
   l1(2)=atan(tan βxn)/(2*pi);
   if ( l1(2) < 0 )
      l1(2)=l1(2) + 0.5;
   end
   disp('There are two solutions.')
      sprintf('The series transmission line can be %.05g wavelengths
long.',l1(1))
   zp=zo*(yo+j*yl*tan βxp)/(yl+j*yo*tan βxp);
   x=-1*imag(zp);
   l2(1)=-1*atan(x*yo)/(2*pi);
   if ( l2(1) < 0 )
      l2(1)=l2(1)+0.5;
   end
   sprintf('Use a shorted series line of %0.5g wavelengths long.',l2(1))
   l2(2)=acot(yo*x)/(2*pi);
   if ( l2(2) < 0 )
      l2(2)=l2(2)+0.5;
   end
   sprintf('Or, use an open series line of %0.5g wavelengths long.',l2(2))
      sprintf('Solution two is a series transmission line %.05g wavelengths
long.',l1(2))
   zp=zo*(yo+j*yl*tan βxn)/(yl+j*yo*tan βxn);
   x=-1*imag(zp);
   l2(3)=-1*atan(x*yo)/(2*pi);
   if ( l2(3) < 0 )
      l2(3)=l2(3)+0.5;
   end
   sprintf('Use a shorted series line of %0.5g wavelengths long.',l2(3))
   l2(4)=acot(yo*x)/(2*pi);
   if ( l2(4) < 0 )
      l2(4)=l2(4)+0.5;
   end
   sprintf('Or, use an open series line of %0.5g wavelengths long.',l2(4))
end
```

6

Amplifier Design

6.1   Introduction

This chapter focuses on microwave amplifier circuit design. A simple microwave amplifier consists of an active device, such as a transistor, an input- and output-matching circuit, and a power supply bias circuit. We begin with the two-port S-parameters for RF and microwave transistors and an operating environment for the finished amplifier. For example, we can design a circuit that amplifies signals from a generator with a 50 ohm impedance and deliver these amplified signals to a 50 ohm load. The transistor S-parameters along with some mapping functions are used to determine the correct impedance, which will terminate the input and output of the transistor. Matching circuits are designed to transform these impedances to the actual source and load impedances where the amplifier is to be used. We will use the circuit synthesis methods discussed in Chapter 5 to build input- and output-matching circuits.

Section 6.2 explains how S-parameters are presented in transistor data sheets. The S-parameters are different for each transistor and depend on how the transistor is packaged. Furthermore, the S-parameters vary with frequency and bias point. Several tables of S-parameters in the manufacturer's data sheets indicate a representative bias voltage and current. Also listed are the magnitude and phase of the two-port S-parameters at different frequencies.

Section 6.3 introduces the concept of amplifier stability. We will be working with transistors that amplify signals over a wide frequency range and are potentially unstable. An unstable transistor can oscillate and make the circuit useless as an amplifier. We will demonstrate how to analyze the S-parameters for possible instabilities in the amplifier design. The analysis will result in regions of the input and output impedance plane that are forbidden for a stable amplifier. A matching circuit that presents these impedances to the transistor should not be designed.

Section 6.4 discusses how to find the optimum matching circuit impedance for a transistor. As shown in Chapter 5, a matching circuit transforms one impedance to another. An amplifier input- or output-matching circuit transforms the characteristic impedance of a transmission line (50 ohms) to an impedance that matches the transistor's input and output.

Section 6.5 defines the different forms of power gain. We have seen that microwave circuit parameters and analysis are functions of power. Power gain is the premier characteristic of high-frequency amplifiers. This section introduces several ways to express the power gain of a microwave amplifier.

Section 6.6 shows how to apply DC power to the amplifier. We must be able to bias the transistor without affecting the high-frequency part of the circuit. This last section introduces a few simple ways of setting the proper voltage and current in a transistor and explains how to apply the power in a way that it will not influence the high-frequency performance of the amplifier.

6.2   Transistor Data Sheets

Transistor manufacturers publish the two-port S-parameters of their transistors in data sheets, which help to predict their microwave properties. The two-port S-parameters can be found in data sheets or on computer disks, or they can be calculated using a transistor model supplied by the manufacturer. Sometimes, especially for UHF and VHF transistors, the Y-parameters or hybrid parameters are given. Various factors, including production variation, bias point, frequency, packaging, and temperature, can affect the S-parameters of a transistor. This section shows how the S-parameters and other data given in data sheets can be used to evaluate how well a particular transistor will work.

Table 6-1 shows an example of two-port S-parameters as they are presented in the data sheet. The S-parameters of the transistor change over frequency and are usually furnished in a table. This particular transistor is a field effect transistor (FET). Its parameters are given in a common-source circuit configuration, which means that the source of the FET is grounded, the gate is attached to Port 1, and the drain is Port 2, as shown in Figure 6-1. Table 6-1 shows the S-parameters listed as magnitude and phase of S_{11}, S_{21}, S_{12}, and S_{22} at discrete frequencies. The S-parameters will vary from part to part due to production variations. The S-parameters listed in the data sheets are usually a statistical mean over many production units.

Table 6-1 *S-parameters of a NE76000 with V_{ds} = 3 V and I_{ds} = 10 mA.*

Freq. GHz	Mag. S_{11}	Ang. S_{11}	Mag. S_{21}	Ang. S_{21}	Mag. S_{12}	Ang. S_{12}	Mag. S_{22}	Ang. S_{22}
0.5	0.99	-7	3.28	175	0.009	85	0.68	-3
1.0	0.99	-14	3.19	169	0.02	81	0.67	-8
2.0	0.99	-27	3.19	158	0.04	74	0.67	-16
4.0	0.95	-50	2.95	138	0.07	59	0.69	-30
6.0	0.89	-70	2.67	120	0.09	47	0.60	-42
8.0	0.86	-87	2.45	104	0.11	36	0.58	-53
10.0	0.81	-104	2.24	90	0.12	29	0.57	-63
14.0	0.74	-135	1.93	63	0.13	12	0.55	-81
18.0	0.70	-155	1.65	43	0.13	3	0.55	-94
22.0	0.67	-171	1.46	26	0.12	4	0.55	-107
26.0	0.26	164	1.29	7	0.12	04	0.55	-123

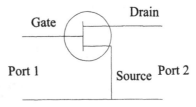

Figure 6-1 *The transistor connected in a common-source configuration.*

These S-parameters are given for the transistor in chip form. Transistors come in several different packages. NEC sells the transistor above in an unpacked version (NE76000) or in a ceramic package (NE76084). Also, notice that the transistor bias point is shown in the title of the table. In this case, the S-parameters are given for the transistor being operated with a 5 volt drain-to-source voltage and a drain current of 10 mA. Keep in mind that the S-parameters of a transistor will differ at various bias points.

Sometimes additional parameters are shown in the data sheets. They can usually be calculated from the S-parameters and are summarized in the table for your convenience. The remainder of this chapter describes these parameters and show how to calculate them from the two-port S-parameters. One of these parameters is the stability factor *K*. It is a measure of how stable the transistor is when placed in different circuits. You may also find noted maximum stable gain (MSG) and maximum available gain (MAG), which are both measures of how much signal gain can be obtained from the transistor at the different frequencies.

6.3 Transistor Stability

The stability of a transistor or amplifier circuit is a measure of the likelihood of oscillation when the circuit is energized. However, the conditions that cause oscillation are more stringent than the conditions that make a circuit unstable. Evidence of instability, or possible instability, can be seen by analyzing the reflection coefficient of a network. Consider a situation in which we have a transistor with the output-terminated with some circuit or load. We will look at the reflection coefficient at the input of the transistor either by calculation or measurement. If the magnitude of the reflection coefficient is greater than 1, more power will be reflected from the circuit than is incident upon it. This phenomenon is called *reflection gain*. With this combination of transistor and load, the circuit could begin to oscillate. By way of the parameter K, we can show how likely it is that the reflection coefficient will be greater than 1, using all possible ways of terminating the transistor.

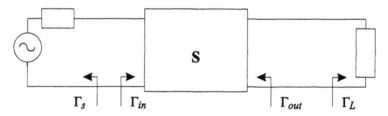

Figure 6-2 *A transistor two-port.*

First we will study the conditions that might cause the input reflection coefficient of a transistor to be greater than 1. Armed with the mapping functions from Chapter 3, we can easily find all the possible output circuits that might cause the magnitude of the input reflection coefficient to be greater than 1. The circuit in Figure 6-2 shows a simple circuit diagram with the input and output reflection coefficient of a transistor two-port. It might seem backwards to map the Γ_{in} plane onto the Γ_L plane but, as will be shown, it will be to our advantage to begin this way. We map the $|\Gamma_{in}| = 1$ circle through the transistor to determine where that circle lies on the Γ_{out} plane. Then we will know all the possible combinations of Γ_L that will cause Γ_{in} to be equal to or greater than 1.

The $|\Gamma_{in}| = 1$ circle is centered at the origin and has a radius of 1. Equation 4.28 gives the input reflection coefficient as a function of the two-port S-parameters and the output termination

$$\Gamma_{in} = \frac{S_{11} - \Delta\Gamma_L}{(1 - S_{22}\Gamma_L)} \qquad\qquad 6.1$$

where $\Delta = S_{11}S_{22} - S_{12}S_{21}$. After we rearrange terms, we arrive at a mapping function relating Γ_L to Γ_{in}

$$\Gamma_L = \frac{\Gamma_{in} - S_{11}}{(S_{22}\Gamma_{in} - \Delta)} \qquad 6.2$$

Using Equation 4.24, we map the $|\Gamma_{in}| = 1$ circle on the Γ_L plane with a center at

$$C_L = \frac{S_{22}* - \Delta*S_{11}}{|S_{22}|^2 - |\Delta|^2} \qquad 6.3$$

and a radius of

$$R_L = \left| \frac{S_{12}S_{21}}{|S_{22}|^2 - |\Delta|^2} \right| \qquad 6.4$$

This defines a circle on the Γ_L plane where a load reflection coefficient (Γ_L) would cause $|\Gamma_{in}| = 1$. This is called the *input stability circle*. If Γ_L were somewhere on this circle, the input reflection coefficient would be equal to 1. This defines the boundary between the two regions of the load reflection coefficient plane. In one region the input reflection coefficient will always be less than 1. Another territory also causes the input reflection coefficient to be equal to or greater than 1, and it causes the two-port to be potentially unstable. We can now determine whether the values inside or outside the circle cause $|\Gamma_{in}|$ to be greater than 1. The easiest way to do this is to test the point at the center of the Smith chart where $\Gamma_L = 0$.

$$\Gamma_{in} = S_{11}|_{\Gamma_L = 0} \qquad 6.5$$

If $|S_{11}|$ is less than 1, the $\Gamma_L = 0$ point lies in the region of stability. If $|S_{11}|$ is greater than 1, the $\Gamma_L = 0$ point is in the region of potential instability.

In addition, we must also know when a source impedance causes the output reflection coefficient to be greater than 1, which would also be potentially unstable. When the input is terminated with a source impedance in a reflection coefficient of Γ_s, the reflection coefficient looking into the two-port is

$$\Gamma_{out} = \frac{S_{22} - \Delta\Gamma_s}{1 - S_{11}\Gamma_s} \qquad 6.6$$

Rearranging terms, we can express Γ_s as a function of the two-port S-parameters and Γ_{out}.

$$\Gamma_s = \frac{S_{22} - \Delta\Gamma_{out}}{(1 - S_{11}\Gamma_{out})} \qquad 6.7$$

Using Equation 4.24, we map the $|\Gamma_{out}| = 1$ circle on the Γ_s plane with a center at

$$C_s = \frac{S_{11}{}^* - \Delta{}^* S_{22}}{|S_{11}|^2 - |\Delta|^2} \qquad 6.8$$

and a radius of

$$R_s = \left| \frac{S_{12}S_{21}}{|S_{11}|^2 - |\Delta|^2} \right| \qquad 6.9$$

This *output stability circle* defines all the possible source reflection coefficients that cause the output reflection coefficient to be equal to 1. Again, we have two regions, one inside the circle and one outside. The point where $\Gamma_s = 0$ is used as a test to see which region is potentially unstable.

$$\Gamma_{out} = S_{22}\big|_{\Gamma_s = 0} \qquad 6.10$$

When S_{11} is less than 1, the region where $\Gamma_s = 0$ is in the stable regions and when $S_{11} > 1$, the region that includes $\Gamma_s = 0$ is the potentially unstable region of the Γ_s plane.

Example 6.1: Find the source and load stability circles for the NE76000 at 8 GHz. According to Table 6.1, the two-port S-parameters are

$S_{11} = 0.86@{-}87° \quad S_{21} = 2.45@104° \quad S_{12} = 0.11@36° \quad S_{22} = 0.58@{-}53°$

$\Delta = (0.86@{-}87°)\,(0.58@{-}53°) - (2.45@104°)\,(0.11@36°) = .5264@{-}109.4°$

The source stability circle is given by Equations 6.8 and 6.9.

$$C_s = \frac{(0.86@{-}87°){}^* - (.5264@{-}109.4°){}^* (0.58@{-}53°)}{(0.86)^2 - (0.5264)^2}$$
$$= 1.334@101.6°$$

$$R_s = \left| \frac{(2.45)(0.11)}{|0.86|^2 - |0.5264|^2} \right|$$
$$= 0.583$$

The source stability circle is a collection of all the points on the source reflection coefficient plane that cause the output reflection coefficient to be equal to 1 and is shown in Figure 6-3. Since $|S_{22}| < 1$, all points inside this circle will cause the output reflection coefficient to be greater than 1. This region is potentially unstable. The load stability circle is calculated using Equations 6.3 and 6.4.

$$C_L = \frac{(0.58@-53°)* - (.5264@-109.4°)*(0.86@-87°)}{|0.86|^2 - |0.5264|^2}$$

$$= 5.0415@103.4°$$

$$R_L = \left| \frac{(2.45)(.11)}{|0.58|^2 - |0.5264|^2} \right|$$

$$= 4.5477$$

The load stability circle is also shown in Figure 6-3. Since $|S_{11}| < 1$, all points inside the load stability circle are potentially unstable.

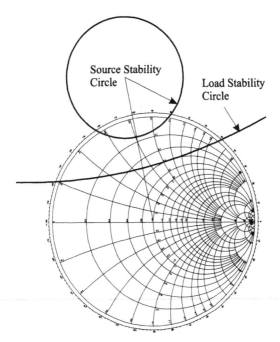

Figure 6-3 *The source and load stability circles plotted on a Smith chart.*

When the load and source stability circle are *both* outside the $\Gamma = 1$ circle on the Smith chart, the transistor is unconditionally stable. This means that there is no combination of passive source or load termination that can cause the transistor to oscillate. Or, stated mathematically, if $|\Gamma_s| <= 1$ and $|\Gamma_L| <= 1$, then $|\Gamma_{in}| < 1$ and $|\Gamma_{out}| < 1$. If this is not the case, the transistor is conditionally stable.

If the radius of the stability circle is less than 1 plus the magnitude of the center of the circle, the circle will always stay outside the Smith chart. As long as the center of the Smith chart is stable, there will be no passive impedance in the unstable region. For the source impedance, the condition of unconditional stability is satisfied when

$$|\,|\,C_S| - R_S| > 1 \text{ and } |S_{22}| < 1 \qquad 6.11$$

For unconditional stability, the load stability circles must also satisfy

$$|\,|C_L| - R_L| > 1 \text{ and } |S_{11}| < 1 \qquad 6.12$$

There are four conditions that must be satisfied for the two-port to be unconditionally stable. If $|S_{11}|$ or $|S_{22}|$ is greater than 1, the two-port is potentially unstable when Γ_L or Γ_s is equal to zero. It is more convenient to rewrite the conditions as follows. When

$$|\Gamma_s| < 1 \qquad 6.13$$

and

$$|\Gamma_L| < 1 \qquad 6.14$$

then

$$|\Gamma_{in}| = \left| \frac{S_{11} - \Delta\Gamma_L}{1 - S_{22}\Gamma_L} \right| < 1 \qquad 6.15$$

and

$$|\Gamma_{out}| = \left| \frac{S_{22} - \Delta\Gamma_s}{1 - S_{11}\Gamma_s} \right| < 1 \qquad 6.16$$

for the two-port to be unconditionally stable. Both sides of Equations 6.15 and 6.16 are squared, and we collect like terms.

$$|S_{11}|^2 + |\Delta|^2|\Gamma_L|^2 - 2\operatorname{Re}\{S_{11}\Delta\Gamma_L\} < 1 + |S_{22}|^2|\Gamma_L|^2 - 2\operatorname{Re}\{S_{22}\Gamma_L\} \qquad 6.17$$

$$|S_{22}|^2 + |\Delta|^2|\Gamma_s|^2 - 2\text{Re}\{S_{22}\Delta\Gamma_s\} < 1 + |S_{11}|^2|\Gamma_s|^2 - 2\text{Re}\{S_{11}\Gamma_s\} \qquad 6.18$$

Since $|S_{11}| < 1$, $|S_{22}| < 1$, $|\Gamma_s| < 1$, and $|\Gamma_L| < 1$, we arrive at the expression

$$2|S_{12}S_{21}| < 1 - |S_{11}|^2 - |S_{22}|^2 + |\Delta|^2 \qquad 6.19$$

as long as $|\Delta| < 1$. We now can define a formula for K that must be greater than 1 if the transistor is unconditionally stable

$$K = \frac{1 - |S_{11}|^2 - |S_{22}|^2 + |\Delta|^2}{2|S_{12}S_{21}|} > 1 \qquad 6.20$$

as long as the origin of the source reflection coefficient plane is stable, which results in one more test.

$$|\Delta| < 1 \qquad 6.21$$

Equations 6.20 and 6.21 are necessary and sufficient conditions for unconditional stability. If K is less than 1 or $|\Delta| > 1$, the transistor is conditionally stable, and the stability circles should be calculated. If K is greater than 1 and $|\Delta| < 1$, the transistor is unconditionally stable, and we do not need to investigate the stability of the device.

Example 6.2: Find the stability factor K for the NE76000 in Table 6-1 at 8 GHz. We have already calculated Δ in Example 6.1.

$$|\Delta| = .5264 < 1$$

We need to calculate K.

$$K = \frac{1 - 0.86^2 - 0.58^2 + .5264^2}{2(2.45)(0.11)} = 0.3731 < 1$$

This FET is only conditionally stable at 8 GHz. This agrees with the results of Example 6.1 showing a source stability circle that encompasses some of the Smith chart.

6.4   Conjugate Matching Circuits

So far we have shown how to interpret the data sheets and find the regions of input and output reflection coefficient to design a stable amplifier. Now we can find an analytic way to design the input- and output-matching circuits of an amplifier. If the center of the Smith chart is stable, we can simply connect the transistor with 50 ohm transmission lines. The power gain of the circuit will be equal to

$$\left|\frac{b_2}{a_1}\right|^2 = |S_{21}|^2$$

6.22

However, this will probably not be the largest gain that a transistor could have provided the circuit is optimized. Given the cost of microwave transistors, we want to obtain as much gain as possible. Thus, the transistor has be matched. Matching will transform the input and output impedance of the device to 50 ohms. The derivation of the stability circles has shown that the input match generally affects the output impedance of the device and vice versa. If we know what load or source impedance is connected to the device, the other matching circuit can be created. However, the maximum possible gain is obtained when both matching circuits are designed to achieve a simultaneous match of both input and output under the condition of maximum gain.

Equation 6.1 gives the input reflection coefficient of a two-port when Port 2 is terminated in an impedance with a reflection coefficient of Γ_L. Equation 6.6 shows the output reflection coefficient of a two-port with Port 1 terminated with Γ_s. We will force the input reflection coefficient given in Equation 6.1 to be equal to the conjugate of the source reflection coefficient, i.e., $\Gamma_{in} = \Gamma_s{}^*$. Simultaneously, we will make the output reflection coefficient equal to the conjugate of the load reflection coefficient.

$$\Gamma_{out} = \Gamma_L^* = \frac{S_{22} - \Delta\Gamma_s}{1 - S_{11}\Gamma_s}$$

6.23

This equation for Γ_L is substituted into Equation 6.1.

$$\Gamma_{in} = \Gamma_s^* = \frac{S_{11} - \Delta\left(\dfrac{S_{22} - \Delta\Gamma_s}{1 - S_{11}\Gamma_s}\right)^*}{1 - S_{22}\left(\dfrac{S_{22} - \Delta\Gamma_s}{1 - S_{11}\Gamma_s}\right)^*}$$

6.24

$$\Gamma_s = \frac{S_{11}^* - |S_{11}|^2\Gamma_s - \Delta^* S_{22} + |\Delta|^2\Gamma_s}{1 - S_{11}\Gamma_s - |S_{22}|^2 + S_{22}^*\Delta\Gamma_s}$$

6.25

$$\Gamma_s - S_{11}\Gamma_s^2 - |S_{22}|^2\,\Gamma_s + S_{22}^*\Delta\Gamma_s^2 = S_{11}^* - |S_{11}|^2\,\Gamma_s - \Delta^* S_{22} + |\Delta|^2\,\Gamma_s \qquad 6.26$$

Solving Equation 6.25 for Γ_s gives us the source conjugate match

$$\Gamma_{sM} = C_1^* \left[\frac{B_1 \pm \sqrt{B_1^{\,2} - 4|C_1|^2}}{2|C_1|^2} \right] \qquad 6.27$$

where

$$C_1 = S_{11} - \Delta S_{22}^* \qquad 6.28$$

and

$$B_1 = 1 + |S_{11}|^2 - |S_{22}|^2 - |\Delta|^2 \qquad 6.29$$

We will chose the sign in Equation 6.27 so that $|G_{sM}| < 1$ when the device is unconditionally stable. The square root in Equation 6.27 can be written as

$$\frac{1}{B_1} \sqrt{1 - \left(\frac{2|C_1|}{B_1}\right)^2} = 1 - \frac{1}{2}\frac{(2|C_1|)^2}{B_1} - \frac{1}{8}\frac{(2|C_1|)^4}{(B_1)^3} - \frac{1}{16}\frac{(2|C_1|)^6}{(B_1)^5}\;\ldots \qquad 6.30$$

When $B_1 < 0$, Equation 6.29 is likely to be a large negative number; thus, we use the plus sign when $B_1 < 0$. Conversely, we choose the minus sign when $B_1 > 0$. Repeating the same process but solving for the load conjugate match, we find that

$$\Gamma_{LM} = C_2^* \left[\frac{B_2 \pm \sqrt{B_2^{\,2} - 4|C_2|^2}}{2|C_2|^2} \right] \qquad 6.31$$

where

$$C_2 = S_{22} - \Delta S_{11}^* \qquad 6.32$$

and

$$B_2 = 1 - |S_{11}|^2 + |S_{22}|^2 - |\Delta|^2 \qquad 6.33$$

Example 6.3: A Motorola MRF951 RF transistor has the following S-parameters at 2.5 GHz at the bias point of 6 volts from collector to emitter and a collector current of 5 mA.

$$S_{11} = 0.56@140° \quad S_{21} = 1.74@32° \quad S_{12} = 0.15@48° \quad S_{22} = 0.41@-92°$$

Find the conjugate match on the input and output and design the matching circuits for the amplifier. First, we check to see if the transistor is unconditionally stable.

$$\Delta = (0.56@140°)\,(0.41@-92°) - (1.74@32°)\,(0.15@48°) = .1386@-38.6° < 1$$

$$K = \frac{1 - (0.56)^2 - (.41)^2 + (.1386)^2}{2*(1.74)(0.15)} = 1.0297$$

The transistor is stable at this frequency and bias point. We can calculate the simultaneous conjugate match.

$$B_1 = 1 + |0.56|^2 - |0.41|^2 - |.1386|^2 = 1.1263$$

$$B_2 = 1 + |0.41|^2 - |0.41|^2 - |.1386|^2 = 0.8353$$

$$C_1 = 0.56@140° - (.1386@-38.6°)\,(0.41@-92°)^* = 0.5595@146°$$

$$C_2 = 0.41@-92° - (.1386@-38.6°)\,(0.56@140°)^* = 0.5595@-81°$$

Since both B_1 and B_2 are greater than zero, the minus sign is used in Equations 6.27 and 6.31.

$$\Gamma_{sM} = (0.560@146) * \left[\frac{(1.126) - \sqrt{(1.126)^2 - 4(0.560)^2}}{2(0.560)^2}\right] = 0.892@-145.8°$$

$$\Gamma_{LM} = (0.413@-81) * \left[\frac{(0.835) - \sqrt{(0.835)^2 - 4(0.413)^2}}{2(0.413)^2}\right] = 0.857@81.2°$$

The circuit synthesis technique of a series transmission line and open shunt stub is chosen for the input and output circuit. A circuit is designed to match the reflection coefficient of Γ_{sm}^* on the input and the output circuit is designed to match a reflection coefficient of Γ_{Lm}^*.

$$Z_{in} = 50\left(\frac{1 + \Gamma_{sm}^*}{1 - \Gamma_{sm}^*}\right) = 3.122 + j15.32$$

$$Z_{out} = 50 \left(\frac{1 + \Gamma_{Lm}^{*}}{1 - \Gamma_{Lm}^{*}} \right) = 9.039 - j57.54$$

The input and output impedances are plotted on an impedance Smith chart in Figure 6-4. Both impedances are rotated around the Smith chart clockwise (toward the generator) until they lie on the constant 0.02 mho conductance circle. The length of the open transmission lines on either end cancels the imaginary part of the admittance. A diagram of the circuit is shown in Figure 6-5. Figure 6-6 contains a plot of the simulated input and output reflection coefficient and gain of the amplifier.

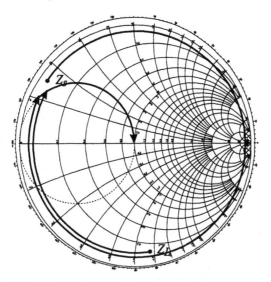

Figure 6-4 *The input and output matching circuit design plotted on a Smith chart.*

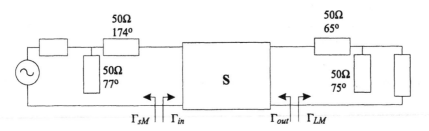

Figure 6-5 *The amplifier circuit for Example 6.3 using transmission lines in a single-stub matching circuit topology on the input and output.*

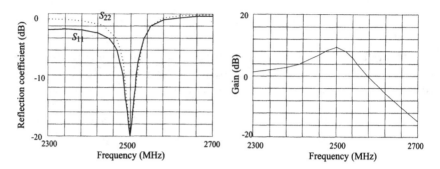

Figure 6-6 *The input and output reflection coefficient of the circuit in Example 6.3 is plotted in (a) and the gain is plotted in (b), as modeled by =SuperStar=.*

This example illustrates how a poor choice of circuit topology affects the performance of an amplifier. The circuit is very narrowband and will be susceptible to minor manufacturing variations. The long series transmission line on the input is very frequency-sensitive. A lumped-element or double-stub topology could be a better choice for this transistor at this frequency.

Example 6.4: The Fujitsu FHX04/LG FET has the following S-parameters at 12 GHz at the bias point of 2 volts from drain to source and a drain current of 10 mA.

$$S_{11} = 0.601@178.5° \quad S_{21} = 2.245@-5.7° \quad S_{12} = 0.076@-15.9°$$
$$S_{22} = 0.587@-146.4°$$

Solve for the conjugate match for this transistor. Again, we must check the stability of the transistor at this frequency and bias point.

$$\Delta = (0.601@178.5°)(0.587@-146.4°) - (2.245@-5.7°)(0.076@-15.9°)$$
$$= .2869@60.74° < 1$$

$$K = \frac{1 - (0.601)^2 - (.587)^2 + (.2869)^2}{2*(2.245)(0.076)} = 1.1035$$

The transistor is unconditionally stable. We can continue by finding B_1, B_2, C_1, and C_2.

$$B_1 = 1 + |0.601|^2 - |0.587|^2 - |.2869|^2 = 0.934$$

$$B_2 = 1 + |0.587|^2 - |0.601|^2 - |.2869|^2 = 0.901$$

$$C_1 = 0.601@178.5° - (.286@60.74°)(0.587@-146.4°)* = 0.4603@168.4°$$

$$C_2 = 0.587@146.4° - (.286@60.74°)(0.601@-178.5°)* = 0.4435@157.1°$$

Since both B_1 and B_2 are greater than 0, the minus sign is used in Equations 6.27 and 6.31.

$$\Gamma_{sM} = (0.4603@168.4)*\left[\frac{(0.934) - \sqrt{(0.634)^2 - 4(0.4603)^2}}{2(0.4603)^2}\right] = 0.842@-168.4°$$

$$\Gamma_{LM} = (0.4435@-157.1)*\left[\frac{(0.901) - \sqrt{(0.901)^2 - 4(0.4435)^2}}{2(0.4435)^2}\right] = 0.837@157.14°$$

The input and output matching circuits need to be designed to match to an input and output impedance of

$$Z_{in} = 4.33 + j5.04$$

$$Z_{in} = 4.63 - j10.03$$

The design of the input and output matching circuits is shown on the Smith chart in Figure 6-7. A simplified schematic of the amplifier is shown in Figure 6-8, as implemented using single-shunt open stub circuits. The input and output reflection coefficient and gain shown were modeled by =SuperStar=.

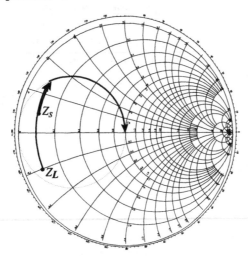

Figure 6-7 *The Smith chart showing the design of the matching circuits in Example 6.4. Input and output impedance are both rotated to the top of the constant Y_o circle, at which point a shunt stub cancels the imaginary part.*

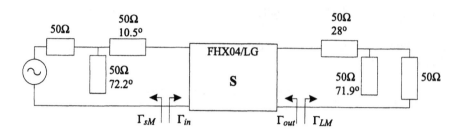

Figure 6-8 *The amplifier circuit in Example 6.4.*

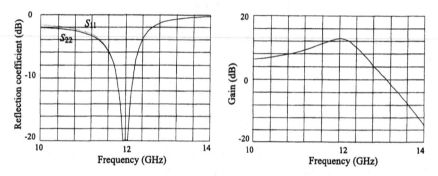

Figure 6-9 *The input and output reflection coefficient and gain of the amplifier.*

The conjugate match cannot be used if it falls in an unstable region of the Smith chart. If there is no sensible conjugate match when the transistor is only conditionally stable, another point must be chosen. If the transistor is unconditionally stable, the input and output conjugate matching circuit can be synthesized using the techniques discussed in Chapter 5. If lossless matching circuits are used, the amplifier will have the maximum stable gain. However, if the conjugate match is inside an unstable region, we have to chose some other matching point. The gain will vary significantly, depending on the choice of the match point.

6.5 Power Gain

This section describes the different forms of gain and how they are calculated. If the conjugate match is not stable, it could take some time to choose a match by trial and error that results in a reasonable gain. For each try, a matching circuit would have to be designed and entered into the microwave circuit simulator. It is more efficient to use the mapping functions from Chapter 4 and calculate circles of constant gain. There are a several different ways to measure amplifier gain.

Figure 6-10 shows a signal flow diagram of a source and load. The power delivered to a load is the incident power minus the reflected power.

$$P = |a|^2 - |b|^2 \qquad 6.34$$

The power delivered to a load by a transmission line with a perfectly matched source is

$$P_L = \frac{P_{in}}{1 - |\Gamma_L|^2} \qquad 6.35$$

When the source is not matched, the power delivered to a load is

$$P = \frac{1}{|1 - \Gamma_s \Gamma_L|^2} \qquad 6.36$$

When an amplifier is inserted between a generator and a load, it is important to know how much the signals will be amplified. There could be reflections from the load, the generator, or the input and output of the two-port. This creates an opportunity to describe the gain of the amplifier in several different ways.

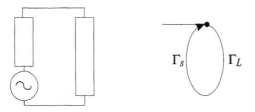

Figure 6-10 *A generator with a reflection coefficient of Γ_s connected to a load with a reflection coefficient Γ_L.*

Power gain is usually expressed by one of the following three quantities: transducer power gain (G_t), power gain, or operating power gain, (G_p) and available power gain (G_a). Transducer power gain is the ratio of power delivered to the load to the power available from the source.

$$G_t = \frac{P_L}{P_{aval}} = \frac{\text{power delivered to the load}}{\text{power available from the source}} \qquad 6.37$$

Transducer power gain is the most general of the three quantities and is derived by finding the power gain lost at the input of the two-port, the gain through the two-port, and the gain lost by not delivering all the power to the load. The power gain available from any source is

$$|a_s|^2\left(1-|\Gamma_s|^2\right)$$

6.38

The power that is delivered to the load is

$$|b_2|^2\left(1-|\Gamma_L|^2\right)$$

6.39

The transducer power gain is

$$G_t = \frac{|b_2|^2}{|a_s|^2}\left(1-|\Gamma_L|^2\right)\left(1-|\Gamma_s|^2\right)$$

6.40

Solving for a_s and b_2

$$\frac{b_2}{a_1} = \frac{S_{21}}{1-S_{22}\Gamma_L}$$

6.41

and

$$\frac{a_1}{a_s} = \frac{1}{1-\Gamma_{in}\Gamma_s}$$

6.42

Substituting these into Equation 6.39, we arrive at

$$G_t = \frac{1-|\Gamma_s|^2}{|1-\Gamma_{in}\Gamma_s|^2}|S_{21}|^2\frac{1-|\Gamma_L|^2}{|1-S_{22}\Gamma_L|^2}$$

6.43

The transducer power gain could have been derived going from input to output, which would have yielded this equivalent formula.

$$G_t = \frac{1-|\Gamma_s|^2}{|1-S_{11}\Gamma_s|^2}|S_{21}|^2\frac{1-|\Gamma_L|^2}{|1-\Gamma_{out}\Gamma_L|^2}$$

6.44

Γ_{in} is the input reflection coefficient of the two-port. It is a function of the two-port S-parameters and the load reflection coefficient, as expressed in Equation 6.1. Substituting Equation 6.1 into Equation 6.42 provides the equation for transducer gain.

$$G_t = \frac{1-\left|\Gamma_s\right|^2}{\left|1 - \dfrac{S_{11} - \Delta\Gamma_L}{(1 - S_{22}\Gamma_L)}\Gamma_s\right|^2}\left|S_{21}\right|^2\frac{1-\left|\Gamma_L\right|^2}{\left|1 - S_{22}\Gamma_L\right|^2} \qquad 6.45$$

If Equation 6.6 is substituted into Equation 6.44, another form of the transducer power gain is obtained.

$$G_t = \frac{1-\left|\Gamma_s\right|^2}{\left|1 - S_{11}\Gamma_s\right|^2}\left|S_{21}\right|^2\frac{1-\left|\Gamma_L\right|^2}{\left|1 - \dfrac{S_{22} - \Delta\Gamma_s}{1 - S_{11}\Gamma_s}\Gamma_L\right|^2} \qquad 6.46$$

Either formula can be used. There are three components to the transducer power gain: gain lost at the input, the power gain of the two-port, and the gain lost at the input. The forward gain of the two-port is $\left|S_{21}\right|^2$, which is reduced by the mismatch at the input

$$\text{Gain lost at the input } = \frac{1-\left|\Gamma_s\right|^2}{\left|1 - S_{11}\Gamma_s\right|^2} \qquad 6.47$$

and by the gain lost by the mismatch at the output.

$$\text{Gain lost at the output } = \frac{1-\left|\Gamma_L\right|^2}{\left|1 - \dfrac{S_{22} - \Delta\Gamma_s}{1 - S_{11}\Gamma_s}\Gamma_L\right|^2} \qquad 6.48$$

Transducer power gain takes into account power lost because of inefficient input and output matching.

The power gain is the ratio of power delivered + 0.* to the load versus the power input to the amplifier network.

$$G_p = \frac{P_L}{P_{in}} = \frac{\text{power delivered to the load}}{\text{power input to the network}} \qquad 6.49$$

It is assumed that the source is perfectly matched to the characteristic impedance of the amplifier. When this condition is inserted into transducer gain, power gain is

$$G_p = \frac{1}{1-\left|\Gamma_{IN}\right|^2}\left|S_{21}\right|^2\frac{1-\left|\Gamma_L\right|^2}{\left|1 - S_{22}\Gamma_L\right|^2} \qquad 6.50$$

We substitute Equation 6.1 into Equation 6.50

$$G_p = \frac{1}{1 - \left|\dfrac{S_{11} - \Delta\Gamma_L}{1 - S_{22}\Gamma_L}\right|^2} |S_{21}|^2 \frac{1 - |\Gamma_L|^2}{|1 - S_{22}\Gamma_L|^2} \qquad 6.51$$

$$G_p = |S_{21}|^2 \frac{1 - |\Gamma_L|^2}{|1 - S_{22}\Gamma_L|^2 - |S_{11} - \Delta\Gamma_L|^2} \qquad 6.52$$

$$G_p = \frac{\left(1 - |\Gamma_L|^2\right)|S_{21}|^2}{1 - |S_{22}|^2 + |\Gamma_L|^2\left(|S_{22}|^2 - |\Delta|^2\right) - 2\,\mathrm{Re}\{\Gamma_L C_2\}} \qquad 6.53$$

where C_2 is given in Equation 6.32. The meaning of power gain can be interpreted by examining the different parts of Equation 6.50. The gain of the two-port $|S_{21}|^2$ is reduced by the power lost by the combined mismatch of the load and the two-port output

$$\text{Gain lost at the output} = \frac{1 - |\Gamma_s|^2}{|1 - S_{22}\Gamma_L|^2} \qquad 6.54$$

and increased because the value S_{21} only gives us the gain when the two-port is connected to a 50 ohm source and load.

$$\text{Gain adjustment at the input} = \frac{1}{1 - |\Gamma_{in}|^2} \qquad 6.55$$

The power gain is sometimes called the operating power gain and is independent of the source reflection coefficient Γ_s.

The available power gain is the ratio of power available at the output of the amplifier to the power available from the source.

$$G_a = \frac{P_{AVN}}{P_{avs}} = \frac{\text{power available from the network}}{\text{power available from the source}} \qquad 6.56$$

For this gain measure, we assume that the load is perfectly matched to the two-port so that all the available power at the output of the two-port is delivered to the load.

$$G_a = \frac{1-|\Gamma_s|^2}{|1-S_{11}\Gamma_s|^2}|S_{21}|^2 \frac{1}{1-|\Gamma_{out}|^2}$$

6.57

Equation 6.6 is substituted into the expression for available power gain

$$G_a = \frac{1-|\Gamma_s|^2}{|1-S_{11}\Gamma_s|^2}|S_{21}|^2 \frac{1}{1-\left|\dfrac{S_{22}-\Delta\Gamma_s}{1-S_{11}\Gamma_s}\right|^2}$$

6.58

$$G_a = |S_{21}|^2 \frac{1-|\Gamma_s|^2}{|1-S_{11}\Gamma_s|^2 - |S_{22}-\Delta\Gamma_s|^2}$$

6.59

$$G_a = \frac{\left(1-|\Gamma_s|^2\right)|S_{21}|^2}{1-|S_{11}|^2 + |\Gamma_s|^2\left(|S_{11}|^2 - |\Delta|^2\right) - 2\operatorname{Re}\{\Gamma_s C_1\}}$$

6.60

where C_1 is given in Equation 6.27. Again, we interpret this as the two-port power gain $|S_{21}|^2$ reduced by the power lost by the source mismatch and increased by the fact that all the power is actually delivered to the load. The available power gain is independent of the load reflection coefficient Γ_L.

Sometimes unilateral transducer power gain (G_{TU}) is used because it simplifies the derivation of some circuit parameters. For the unilateral transducer power gain we assume that the transducer has no reverse transmission, i.e., $|S_{12}| = 0$. When $S_{12} = 0$, $\Gamma_{in} = S_{11}$ turning Equation 6.45 to

$$G_{TU} = \frac{1-|\Gamma_s|^2}{|1-S_{11}\Gamma_s|^2}|S_{21}|^2 \frac{1-|\Gamma_L|^2}{|1-S_{22}\Gamma_L|^2}$$

6.61

Transistor data sheets will often show maximum available gain (MAG) and maximum stable gain (MSG). Maximum available gain is the transducer power gain under conjugate match conditions on the input and output of the transducer. Substituting Γ_{sM} in for Γ_s and Γ_{LM} in for Γ_L into Equation 6.44 yields

$$G_{T\max} = \frac{\left(1-|\Gamma_{sM}|^2\right)|S_{21}|^2\left(1-|\Gamma_{LM}|^2\right)}{\left|(1-S_{11}\Gamma_{sM})(1-S_{22}\Gamma_{LM}) - S_{12}S_{21}\Gamma_{sM}\Gamma_{LM}\right|^2}$$

6.62

which is equal to

$$G_{T\,max} = \frac{|S_{21}|}{|S_{12}|}\left(K - \sqrt{K^2 - 1}\right) \qquad 6.63$$

When $K < 1$, the two-port is potentially unstable. Maximum stable gain is a useful figure of merit in cases where the transistor is only conditionally stable. The maximum stable gain occurs when $K = 1$.

$$MSG = \frac{|S_{21}|}{|S_{12}|} \qquad 6.64$$

Example 6.5: Find G_t, G_p, G_a, G_{TU}, G_{Tmax} and MSG for the amplifier shown in Figure 6-8.

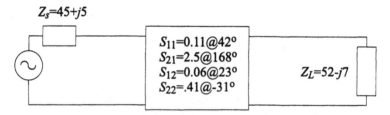

Figure 6-11 *The amplifier circuit for Example 6.5.*

$$\Gamma_s = \frac{45 + j5 - 50}{45 + j5 + 50} = 0.0743@131°$$

$$\Gamma_L = \frac{52 - j7 - 50}{52 - j7 + 50} = 0.0685@-86°$$

$$\Delta = (0.11@42°)(0.41@-31°) - (2.55@168°)(0.06@23°) = 0.195@11°$$

$$\Gamma_{in} = \frac{(0.11@42°) - (0.195@11°)(0.0685@-86°)}{(1 - (0.41@-31°)(0.0685@-86°))} = 0.115@46.4°$$

$$\Gamma_{out} = \frac{(0.41@-31°) - (0.195@11°)(0.0743@131°)}{(1 - (0.11@42°)(0.074@131°))} = 0.421@-31.1°$$

$$G_t = \frac{1 - |\Gamma_s|^2}{|1 - \Gamma_{in}\Gamma_s|^2}|S_{21}|^2\frac{1 - |\Gamma_L|^2}{|1 - S_{22}\Gamma_L|^2} = 5.926$$

$$G_p = \frac{1}{1-|\Gamma_{IN}|^2}|S_{21}|^2 \frac{1-|\Gamma_L|^2}{|1-S_{22}\Gamma_L|^2} = 6.143$$

$$G_{TU} = \frac{1-|\Gamma_s|^2}{|1-S_{11}\Gamma_s|^2}|S_{21}|^2 \frac{1-|\Gamma_L|^2}{|1-S_{22}\Gamma_L|^2} = 5.931$$

$$G_{TU} = \frac{1-|\Gamma_s|^2}{|1-S_{11}\Gamma_s|^2}|S_{21}|^2 \frac{1-|\Gamma_L|^2}{|1-S_{22}\Gamma_L|^2} = 5.931$$

$$G_{T\max} = \frac{|S_{21}|}{|S_{12}|}\left(K - \sqrt{K^2 - 1}\right) = 18.82$$

$$\text{MSG} = 29.54$$

6.6 Constant Gain Circles

We have shown how to calculate the simultaneous conjugate match for a transistor that is unconditionally stable. We can achieve the maximum transducer power gain by matching the input and output of the transistor with circuits that create these reflection coefficients. When the transistor is potentially unstable, there is no safe conjugate matching point. In this case, the gain of the amplifier must be chosen so that it is less than or equal to the maximum stable gain. A systematic method of designing amplifiers with a predictable gain has to be developed. In the case of a potentially unstable transistor, one port must be purposely mismatched because a stable conjugate match does not exist.

First, we will consider the case in which the input is well-matched and the output is purposely mismatched to achieve a specific gain. The power gain in Equation 6.53 is independent of the source impedance and provides a practical method of mismatching the output of a transistor. We have control over the output mismatch that is seen by the transistor, or Γ_L. Contours of constant gain will be in the form of circles on the Γ_L plane. To find these circles, let G_d be the desired gain. We invert Equation 6.53.

$$\frac{1}{G_d} = \frac{1-|S_{22}|^2 + |\Gamma_L|^2\left(|S_{22}|^2 - |\Delta|^2\right) - 2\operatorname{Re}\{\Gamma_L C_2\}}{\left(1-|\Gamma_L|^2\right)|S_{21}|^2} \qquad 6.65$$

We can rearrange Equation 6.65

$$\frac{|S_{21}|^2}{G_d} = -D_2 + \frac{B_2 - 2\operatorname{Re}\{\Gamma_L C_2\}}{1 - |\Gamma_L|^2} \qquad 6.66$$

where

$$D_2 = \left||S_{22}|^2\right| - |\Delta|^2$$

and B_2 and C_2 are given in Equations 6.32 and 6.33. Equation 6.65 can be written as

$$\frac{B_2 - 2\operatorname{Re}\{C_2 \Gamma_L\}}{1 - |\Gamma_L|^2} = \frac{|S_{21}|^2 + G_d D_2}{G_d} = X \qquad 6.67$$

We solve for $|\Gamma_L|$,

$$X^2 |\Gamma_L|^2 - 2X\operatorname{Re}\{C_2 \Gamma_L\} + |C_2|^2 - X^2 + X B_2 - |C_2|^2 = 0 \qquad 6.68$$

which defines circles on the $|\Gamma_L|$ plane with a center at

$$C_p = \frac{G_d\left(S_{22}{}^* - \Delta^* S_{11}\right)}{|S_{21}|^2 + G_d\left(|S_{22}|^2 - |\Delta|^2\right)} = \frac{G_d C_2{}^*}{|S_{21}|^2 + G_d\left(|S_{22}|^2 - |\Delta|^2\right)} \qquad 6.69$$

and a radius of

$$R_p = \frac{|S_{21}|\sqrt{|S_{21}|^2 - 2G_d K |S_{21} S_{12}| + G_d^2 |S_{12}|^2}}{\left||S_{21}|^2 + G_d\left(|S_{22}|^2 - |\Delta|^2\right)\right|} \qquad 6.70$$

These circles are plotted on the output reflection coefficient plane; all the centers will lie at the same angle as $C_2{}^*$. We can plot many of these circles as long as G_d is less than maximum power gain. The circles will show the contours of constant power gain.

Another method of achieving a specific gain is to mismatch the input of the transistor. We use Equation 6.59 for available power gain since it is independent of the load impedance. We can use the same method shown in Equations 6.65 to 6.68 to derive the constant gain circles on the $|\Gamma_s|$ plane. They are centered at

$$C_a = \frac{G_d\left(S_{11}^{\;*} - \Delta^* S_{22}\right)}{|S_{21}|^2 + G_d\left(|S_{11}|^2 - |\Delta|^2\right)} = \frac{G_d C_1^{\;*}}{|S_{21}|^2 + G_d\left(|S_{11}|^2 - |\Delta|^2\right)} \tag{6.71}$$

with a radius of

$$R_a = \frac{|S_{21}|\sqrt{|S_{21}|^2 - 2G_d K|S_{21}S_{12}| + G_d^2|S_{12}|^2}}{\left||S_{21}|^2 + G_d\left(|S_{11}|^2 - |\Delta|^2\right)\right|} \tag{6.72}$$

where C_1 is given in Equation 6.28. Equations 6.71 and 6.72 define circles of constant available power gain that lie on the source reflection coefficient plane.

Example 6.9: Find the input and output constant gain circles for the transistor in Example 6.4.

$$G_{T\,\text{max}} = \frac{|2.245@{-}5.7°|}{|0.076@{-}15.9°|}\left(1.1035 - \sqrt{1.1035^2 - 1}\right) = 18.82$$

The maximum gain is 12.75 dB. Let us plot the 11, 10 and 8 dB gain circles. At 11 dB gain, $G_d = 12.6$. We calculate the constant gain circle of the value 12.6 on the source and load reflection coefficient plane.

$$C_a = \frac{12.6 * 0.4603 \ @{-}168.4°}{|2.245|^2 + 12.6\left(|0.601|^2 - |0.2869|^2\right)} = 0.677@{-}168.4°$$

$$R_a = \frac{|2.245|\sqrt{|2.245|^2 - 2*12.6*1.1035*|2.245*0.076| + 12.6^2|2.245|^2}}{\left||2.245|^2 + 12.6\left(|0.601|^2 - |0.2869|^2\right)\right|} = 0.2894$$

$$C_p = \frac{12.6 * 0.4435@157.1°}{|2.245|^2 + 12.6\left(|0.587|^2 - |0.2869|^2\right)} = 0.6993@157.1°$$

$$R_p = \frac{|2.245|\sqrt{|2.245|^2 - 2*12.6*1.1035|2.245*0.076| + 12.6^2|2.245|^2}}{\left||2.245|^2 + 12.6\left(|0.587|^2 - |0.2869|^2\right)\right|} = 0.2967$$

The other circles are found the same way. The source 10 dB circle has a center at 0.588@–168.4° at a radius of 0.5617, and the load 10 dB circle is centered at 0.5787 at an angle of 157.1 and a radius of 0.5705. The 8 dB circle corresponds to a gain of 6.396. The source constant 8 dB gain circle is centered at 0.4271@–168.4° at a radius of 0.696, and the load 8 dB circle is centered at 0.4179 at an angle of 157.1 and a radius of 0.7033. Figure 6-12 shows the constant gain circles for the source and load plotted on the same Smith chart.

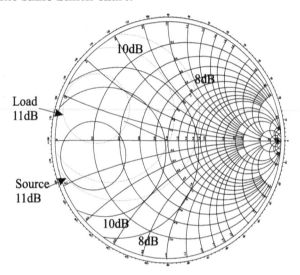

Figure 6-12 *The input and output constant gain circles for the transistor in Example 6.8.*

After the constant gain circles are calculated, we can choose a convenient match that gives us adequate gain. We can then use Equation 6.1 or 6.6 to calculate the match on the other side of the two-port. For example, the source constant 10 dB gain circle in Example 6.9 is shown in Figure 6-12. We choose a convenient source match on the circle and use Equation 6.6 to calculate Γ_{out}. Finally, we design the output matching circuit in order for the two-port to see a load with a reflection coefficient of Γ_{out}^*.

6.7   Transistor Bias Circuits

In this section, we will discuss numerous transistor bias circuits. So far we have only shown how to design the microwave portion of the amplifiers. We have not shown how the low-frequency part of the circuit is designed. Transistors have gain because a small voltage or current at the input of the transistor can control a large output current or voltage. To achieve this, the operating state of the transistor without an input, or quiescent point, must be set properly. The data sheets contain lists of S-parameters, which are only useful when the current and voltage of the transistor match the bias conditions that correspond to those in the table. Usually, a substantial microwave and low-frequency circuit is needed to bias the transistor and make sure that power is supplied to the transistor in such a way that it is transparent to the microwave signal. There are two aspects of the transistor bias network: the bias injection circuit and the bias supply circuit. In this section, we will show ways to connect the power supply with the microwave transmission lines. We will also indicate what voltages are needed from the supply.

In most cases, a method of supplying the bias voltage and current to the transistor must be invisible to the frequencies of interest to the amplifier. Bias injection circuits are needed to connect to the transmission line. However, they should not affect the propagation of high-frequency signals through the transmission line. This is achieved by isolating the high-frequency signals from the power supply. The simplest bias injection circuit is a series inductor. One side of the inductor is attached to the transmission line, the other connects to the power supply. We have to be careful to ensure that the inductor does not create a significant load on the transmission line. The high-frequency impedance of the inductor should be much greater than the characteristic impedance of the transmission line.

When a suitable inductor cannot be found or fabricated, a transmission line structure is common on planar microwave circuits, as shown in Figure 6-13. These circuits consist of a series of high impedance transmission lines and a shunt low-impedance stub. Both transmission lines are a quarter of a wavelength long at the operation frequency of the amplifier. One end of the series transmission line connects to the amplifier circuit and the other to the stub and power supply. There are usually some decoupling capacitors along with the stub. The quarter wavelength open stub creates a virtual short at the end of the series high impedance transmission line. Since the series line is a quarter of a wavelength long, the short is transformed to an open circuit at the amplifier.

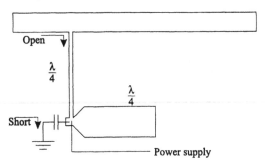

Figure 6-13 *A common transmission line bias injection circuit.*

Once the bias injection networks are designed, the rest of the bias circuit sets the correct voltage and current for the active device. These bias circuits can be either active or passive. Active circuits are used to control the bias of the microwave transistor over temperature and supply voltage. Passive circuits are usually less expensive but also less capable than active circuits. Figure 6-14 shows some popular bias circuits for bipolar transistors. These circuits set the bias point by controlling the emitter current and collector-to-emitter voltage V_{ce}. Figures 6-14(a) and (b) show two methods of biasing a bipolar transistor using a passive bias circuit. The base current of the microwave transistor in Figure 6-14(a) is set by the resistor R_1 and the emitter resistor. The current gain and V_{ce} are controlled by the collector and emitter resistor. In Figure 6-14(b), the resistor dividers R_1 and R_2 help to set the base voltage. The emitter and collector resistors serve the same purpose as they do in the circuit of Figure 6-14(a).

Figures 6-14(c) and (d) show two active bias circuits with a second transistor (T_1), which is used to help bias the microwave transistor and stabilize the collector current over temperature. In the circuit of Figure 6-14(c), T_1 controls the emitter current, and the collector resistor sets V_{ce}. Transistor T_1 in Figure 6-14(d) sets the base current. The current gain and V_{ce} are controlled by the emitter and collector resistors.

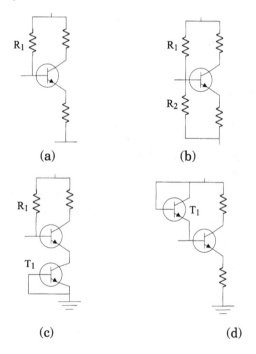

Figure 6-14 *Passive and active bipolar transistor bias circuits.*

Figure 6-15 shows two passive bias circuits for a FET. The gate will usually require a negative gate-to-source voltage (V_{GS}). Figure 6-15(a) shows the use of negative and positive supply voltages to bias the FET. Figure 6-15(b) shows a single supply bias circuit. The voltage dropped across the source resistor creates the negative V_{GS}. Extra care must be exercised to create a high-frequency ground at the source by using capacitors as shown.

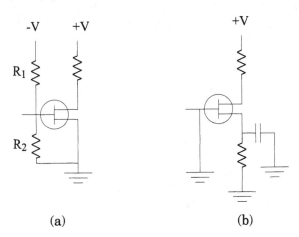

(a)   (b)

Figure 6-15 *Circuits for biasing a Field Effect Transistor (FET).*

6.8 Summary

An active device such as a transistor in an amplifier is necessary to give gain to the circuit. These transistors may need matching circuits because they lack the correct input or output impedance. Some devices are well matched to the characteristic impedance of the transmission line connecting the circuit. However, many transistors need matching circuits at the input and output.

This chapter introduced many useful analysis and design techniques to optimize transistors. The first section discussed information published by manufacturers of high-frequency transistors. The S-parameters are given as two-port magnitude and phase information that are dependent on frequency and bias point. The S-parameters of a transistor vary with the voltage and current of the transistor. Sometimes it is necessary to operate a transistor at various bias points for different applications. The difference between the various bias points are discussed in Chapter 8.

Section 6.3 showed how to calculate the stability of a transistor. Most transistors are potentially unstable at some frequencies. This potential instability is characterized by an input or output reflection coefficient greater than 1. This means that the circuit will exhibit reflection gain, i.e., it will reflect more power than the incident power. Under certain conditions, a growing standing wave can develop and the amplifier can break into oscillations, which will render the circuit useless. Section 6.3 also presented equations for all possible load and source impedances that result in a reflection coefficient greater than 1 at the input or output of the transistor. We can calculate the parameter K to decide if any purely passive load or source impedance causes instability. When K is greater than 1, there is no passive impedance on either the input or the output that can cause the reflection coefficient of the transistor to be greater than 1. These impedances occur in circular regions of the impedance plane and are plotted as stability circles. These regions should be avoided when designing the amplifier's matching circuits.

In previous chapters we have shown that the load impedance of a two-port is likely to affect the input impedance. The output impedance of a two-port is also likely to be affected by the source impedance. When matching a two-port to the characteristic impedance of the transmission lines leading to and from the amplifier, we must simultaneously solve for the load source impedance presented to the transistor. This conjugate match results in a unique source and load impedance to perfectly match a two-port in order to avoid reflection loss at the two-port terminals.

In high-frequency design, it is convenient to work with parameters related to power since amplifier gain has to be expressed in terms of power gain. In this chapter, three forms of power gain were discussed: transducer gain, associated gain and available gain. Transducer power gain takes into account reflection loss at the input and output of the amplifier circuit. Associated power gain makes the power gain independent of the source match. Given a known output match on an amplifier, the associated gain shows how much gain can be obtained given a perfect input match. Available power gain assumes the load is well-matched to the output of the amplifier and describes how much gain will be available given a known input match and a perfect output match.

Sometimes the conjugate match cannot be used. In this case, a systematic method of predicting the gain of an amplifier circuit is needed. Constant gain circles are contours of constant gain plotted on the source or load reflection coefficient plane. They are a valuable tool for designing amplifiers with a specific gain. By plotting these contours, we can choose a specific gain target and design a matching circuit to present that source or load impedance to the transistor.

The last section showed various simple bias circuits. The transistors used should have power supplied at a fixed bias point. Also, the power supply and bias circuit should not interfere with the high-frequency operation of the amplifier circuit. Hence, there is a need for small circuits that inject the bias into the circuit. One example of such a circuit was shown and a simple bias circuit designed to supply the voltage and current to a transistor was briefly explained.

6.9   Problems

6.1   Find the source stability circle for the NE76000 shown in Table 6-1 at 4 GHz.

6.2   Find the source and load stability circles for the NE76000 shown in Table 6-1 at 8 GHz.

6.3   Find the source and load stability circles for the NE76000 shown in Table 6-1 at 4 GHz.

6.4   Find the stability factor K of the Mitsubishi MGF1403 at 6 GHz where the S-parameters are $S_{11} = .933@-108.3°$ $S_{21} = 1.928@72.3°$ $S_{12} = .046@15.5°$ $S_{22} = .724@-81.6°$.

6.5   Find the stability factor K and the stability circles of the Toshiba S8818 at 4 GHz where the S-parameters are $S_{11} = .92@-69°$ $S_{21} = 2.34@112°$ $S_{12} = .047@43°$ $S_{22} = .63@-52°$.

6.6   Find the conjugate source and load match circle for the NE76000 shown in Table 6-1 at 14 GHz if it exists.

6.7   Find the conjugate match and maximum gain for the NEC 76000 shown in Table 6-1 at 8 GHz if it exists.

6.8   Find the maximum stable gain for the transistors in Problems 6.5 and 6.6.

6.9   Calculate the five power gains in a 50 ohm system as shown in Example 6.5 using a source impedance of 75 ohm and a 50 ohm load.

6.10   Calculate the four power gains as illustrated in Example 6.5 using a source impedance of $40 - j20$ ohm and a load impedance of $70 + 10$ ohm.

6.11   Find the 10, 9, 8, 6 dB gain circles on the load reflection coefficient plane for the NE76000 transistor in Table 6-1 at 10 GHz.

6.12 Calculate the maximum transducer gain for the MRF951 given in Example 6.3 and plot the constant 8 dB gain in the source reflection coefficient plane.

6.10   Appendix

MatLab files for calculating stability, conjugate match, and constant gain circles.

stabil.m

```
function[c,r]=stabil(s)
% This function calculates the source and load stability circles
% Usage: [center(2),radius(2)]=stabil(S(2,2))
%    where S(2,2) is the two port S-parameter matrix
%        center(1) is the center of the source stability circle
%        center(2) is the center of the load stability circle
%        radius(1) is the radius of the source stability circle
%        radius(2) is the radius of the load stability circle
j=sqrt(-1);
del=s(1,1)*s(2,2)-s(1,2)*s(2,1);
k=(1-abs(s(1,1))^2-abs(s(2,2))^2+abs(del)^2)/(2*abs(s(1,2)*s(2,1)))
den=abs(s(1,1))^2-abs(del)^2;
c(1)=(conj(s(1,1))-conj(del)*s(2,2))/den;
r(1)=abs(s(1,2)*s(2,1)/den);
den=abs(s(2,2))^2-abs(del)^2;
c(2)=(conj(s(2,2))-conj(del)*s(1,1))/den;
r(2)=abs(s(1,2)*s(2,1)/den);
disp('The center of the source and load circle are:')
c
disp('The radius of the source and load circle are:')
r
```

conjmat.m

```
function[g]=conjmat(s)
% This function calculates the source and load conjugate match
% Usage: [gamma(2)]=conjmat(S(2,2))
%    where S(2,2) is the two port S-parameter matrix
%        gamma(1) is the source conjugate match
%        gamma(2) is the load conjugate match
j=sqrt(-1);
```

```
del=s(1,1)*s(2,2)-s(1,2)*s(2,1);
c1=s(1,1)-del*conj(s(2,2));
b1=1+abs(s(1,1))^2-abs(s(2,2))^2-abs(del)^2;
c2=s(2,2)-del*conj(s(1,1));
b2=1+abs(s(2,2))^2-abs(s(1,1))^2-abs(del)^2;
if ( b1 < 0 )
   temp=b1+sqrt(b1^2-4*abs(c1)^2);
else
   temp=b1-sqrt(b1^2-4*abs(c1)^2);
end
g(1)=conj(c1)*temp/(2*abs(c1)^2);
if ( b2 < 0 )
   temp=b2+sqrt(b2^2-4*abs(c2)^2);
else
   temp=b2-sqrt(b2^2-4*abs(c2)^2);
end
g(2)=conj(c2)*temp/(2*abs(c2)^2);
disp('The source and load conjugate match are:')
g
```

cgain.m

```
function[c,r]=cgain(s,a)
% This function calculates the constant source and load
% gain circles.
% Usage: [c(2,n),r(2,n)]=conjmat(S(2,2),g(n))
%    where S(2,2) is the two port S-parameter matrix
%        g(n) is a vector holding a list of gains
%        n is the number of circles wanted
%        c is the center of the circles
%        r is the radius of the circles
j=sqrt(-1);
del=s(1,1)*s(2,2)-s(1,2)*s(2,1);
k=(1-abs(s(1,1))^2-abs(s(2,2))^2+abs(del)^2)/(2*abs(s(2,1)*s(1,2)));
c1=s(1,1)-del*conj(s(2,2));
b1=1+abs(s(1,1))^2-abs(s(2,2))^2-abs(del)^2;
c2=s(2,2)-del*conj(s(1,1));
b2=1+abs(s(2,2))^2-abs(s(1,1))^2-abs(del)^2;
if ( b1 < 0 )
   temp=b1+sqrt(b1^2-4*abs(c1)^2);
else
   temp=b1-sqrt(b1^2-4*abs(c1)^2);
end
```

```
g(1)=conj(c1)*temp/(2*abs(c1)^2);
if ( b2 < 0 )
   temp=b2+sqrt(b2^2-4*abs(c2)^2);
else
   temp=b2-sqrt(b2^2-4*abs(c2)^2);
end
g(2)=conj(c2)*temp/(2*abs(c2)^2);
disp('source and load conjugate matches are:')
g
den=abs((1-s(1,1)*g(1))*(1-s(2,2)*g(2))-s(1,2)*s(2,1)*g(1)*g(2))^2;
disp('The maximum gain is:');
gtmax=(1-abs(g(1))^2)*abs(s(2,1))^2*(1-abs(g(2))^2)/den
temp=size(a);
n=temp(2);
jj=0;
for ii=1:n
   if a(ii)<gtmax
     jj=jj+1;
     den=abs(s(2,1))^2+a(ii)*(abs(s(1,1))^2-abs(del)^2);
     c(1,jj)=a(ii)*conj(c1)/den;
     temp=abs(s(2,1))^2-2*a(ii)*k*abs(s(2,1)*s(1,2));
     temp=temp+a(ii)^2*abs(s(1,2))^2;
     r(1,jj)=abs(s(2,1))*sqrt(temp)/abs(den);
     den=abs(s(2,1))^2+a(ii)*(abs(s(2,2))^2-abs(del)^2);
     c(2,jj)=a(ii)*conj(c2)/den;
     temp=abs(s(2,1))^2-2*a(ii)*k*abs(s(2,1)*s(1,2));
     temp=temp+a(ii)^2*abs(s(1,2))^2;
     r(2,jj)=abs(s(2,1))*sqrt(temp)/abs(den);
   end
end
```

Example 6.3 circuit

```
CIRCUIT
TLE 1 6 Z=50 L=?75.7 F=2500
TLE 1 2 Z=50 L=?176.32 F=2500
TWO          2          3          0          O=SP          Z=50
F=C:\DEVELOPMENT\EAGLE\SDATA\motorola\mrf9xx\mrf951a.605
TLE 3 4 Z=50 L=?64.9 F=2500
TLE 4 5 Z=50 L=?73.24 F=2500
DEF2P 1 4 TEST
WINDOW
TEST(50)
```

```
'DSP Ck
'SMH S11
'SMH S22
GPH S11 -20 0
GPH S22 -20 0
GPH S21 -20 20
FREQ
SWP 2300 2700 25
```

Example 6.4 circuit

```
CIRCUIT
TLE 1 6 Z=50 L=?72.23 F=12000
TLE 1 2 Z=50 L=?10.53 F=12000
TWO          2          3          0          O=SP          Z=50
F=C:\DEVELOPMENT\EAGLE\SDATA\FUJITSU\FHX04LG.210
TLE 3 4 Z=50 L=?28.04 F=12000
TLE 4 5 Z=50 L=?71.87 F=12000
DEF2P 1 4 TEST
WINDOW
TEST(50)
'DSP Ck
'SMH S11
'SMH S22
GPH S11 -20 0
GPH S22 -20 0
GPH S21 -20 20
FREQ
SWP 10000 14000 41
```

7

Introduction to Broadband Matching

7.1   Introduction

This chapter focuses on broadband matching methods developed for broadband applications and introduces some of the techniques that are effective for broadband amplifier design. The terms *wideband* and *ultrawideband* are used loosely in the amplifier design community. Their definitions are widely accepted, however. The bandwidth factor is defined as

$$BW = \frac{f_2 - f_1}{\sqrt{f_2 f_1}} \qquad\qquad 7.1$$

where f_1 is the lower frequency limit and f_2 is the upper frequency limit. The bandwidth percentage is 100% times the bandwidth factor. Wideband refers to circuits that have a bandwidth factor between 0.50 and 1.0. A circuit with an octave bandwidth has an upper frequency that is twice the lower frequency, or $f_2 = 2f_1$. The bandwidth factor of an octave bandwidth circuit is

$$\frac{2f_1 - f_1}{\sqrt{2f_1^2}} = \frac{1}{\sqrt{2}} \qquad\qquad 7.2$$

or 70.7% bandwidth. This would qualify as a wideband circuit.

An ultrawideband circuit has a bandwidth factor greater than 1.0, or 100%. A circuit with a decade bandwidth could have an upper frequency limit 10 times higher than the lower frequency limit. The bandwidth factor of such a circuit would be $9/\sqrt{10} = 2.85$, or 285%. A circuit with a decade bandwidth is considered an ultrawideband circuit. Most techniques introduced in this chapter are more applicable to wideband circuits rather than to ultrawideband amplifiers.

Section 7.2 shows the approximate circuit model for the transistor. Some broadband amplifier design techniques use the equivalent circuit model of the transistor. The elements of the transistor model can be used as part of the matching circuit. A transistor model has capacitors and inductors that are absorbed into the design of the matching circuit. The transistor becomes a simple resistive load or source, which makes matching less difficult.

Section 7.3 explains the limitation on bandwidth and the desired performance of the matching circuit. As we demand a wider bandwidth and/or a better match across the band, the number of circuit elements in the matching circuit increases.

Section 7.4 introduces an impedance transformer, which is used when the transistor input or output resembles a simple resistor. We can design an impedance transformer that will match this resistance to the characteristic impedance of the system (50 ohms). The impedance transformers shown in this section will use two or more quarter-wave transmission lines to achieve a broadband match.

Section 7.5 discusses the use of standard filter design methods as applied to amplifier matching circuits. The reactive elements of the transistor model are used in the first couple elements of a filter designed to match two impedances in a bandpass, lowpass or highpass filter characteristic. When designing a lumped-element circuit, transmission line structures are substituted for those high-frequency circuits that cannot use capacitors and inductors.

Section 7.6 describes the use of resistive matching. This technique uses lossy elements to broaden the bandwidth of the resonant circuits used to match the transistor. Resistive matching is effective in designing broadband amplifiers for gain or medium power. However, this method should not be used on the input of low noise amplifiers.

Section 7.7 discusses active matching. Transistors are used as matching circuit elements to match the impedance of a transistor to the transmission line. For example, a common-base amplifier has a relatively high input impedance and low output impedance. We can use a transistor in a common-base configuration to match two impedances such as the 50 ohm transmission line to the low impedance of the microwave transistor.

Section 7.8 describes broadband amplifiers that use more than one transistor. These are balanced and distributed amplifiers. Balanced amplifiers use two transistors that are driven by power splitters. Power reflected by the mismatched transistors is canceled by destructive interference in the power splitters. The distributed amplifier is an ultrawideband technique that uses many transistors in parallel as part of a transmission structure. The transistors amplify the waves propagating along the transmission line that connect them.

7.2   Transistor Models

Broadband matching methods require a deeper understanding of the impedance inside the two-port and how it changes with frequency. The synthesis of a broadband matching circuit becomes less problematic when the nature of the impedance is known. The choice of component elements and topology can depend on the type of impedance (resistive, capacitive, and so on) to be matched. The focus of this chapter is on broadband design, which requires knowledge of the transistor model. We will present a general model for a bipolar and field effect transistor. Many active devices, such as tubes, diodes and quantum devices as well as transistors, require broadband matching circuits. However, if and when there is a new device invented that will make transistors obsolete, the information in this chapter will not become outdated. The design techniques described here will still apply if the engineer knows how to match to impedance models that do not depend on the device used.

Transistor models include the electrical properties of the transistor and how the device is attached, or bonded, to the circuit. We must consider the effects of the bond wires and package when designing the matching circuits because they become part of the two-port being matched. Figure 7-1 shows the model for a bipolar transistor that includes bond wires and a package. We can approximate the input impedance of a bipolar transistor by a parallel combination of a resistor and a capacitor with a series inductor. The bond wires that attach the transistor to the package or circuit cause the inductance. Resistance and capacitance are the result of the base-emitter junction of the transistor. The output circuit can be approximated by a parallel current source and resistor combination with a series inductor as shown in Figure 7-1. At low frequencies, the series inductance of the bond wires can be ignored because the inductance is swamped by the resistance and capacitance of the device.

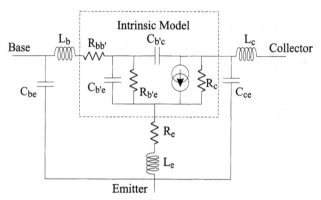

Figure 7-1 *Equivalent model for a bipolar transistor.*

We can find the element values in the model by methods ranging from trial-and-error to sophisticated algorithms. Most manufacturers will supply a model for their transistors. Generating a bilateral model, as the one shown in Figure 7-1, can be difficult. The input impedance depends on the elements on the output, which in turn are affected by the elements on the input. The degree of the interaction depends on the feedback elements of the model. A useful strategy is to determine the resistors and capacitors using the lower frequencies and add the inductors before matching the model with the S-parameters at the higher frequencies of the band.

Frequently, an adequate model can be built using a unilateral model such as the one in Figure 7-2. However, it is not wise to construct a simplified transistor model based on S_{11} and S_{22} alone. A model that approximates the conjugate match of the device and, at least in part, accounts for the forward and reverse gain of the transistor is more useful. Figure 7-2 shows a simplified model for the conjugate match of a NE21938 transistor. Figure 7-3 indicates the input and output conjugate match plotted on the Smith chart as well as S_{11} and S_{22} of the model. The frequency response of the conjugate match will still have a model similar to one based on S_{11} and S_{22} but the element values are somewhat different. A model of the NE21938 (Figure 7-3) was built using the values shown in Table 7-1. The author obtained these values by a trial-and-error method. The model is unlike the one shown in Figure 7-1 but does model the conjugate match of the transistor. Other circuit models that also give the desired S-parameters are possible.

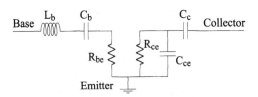

Figure 7-2 *A model simulating the input and output conjugate match of the NE21938 bipolar transistor from 1.5 to 5 GHz at a V_{ce} of 8 volts and a collector current of 8 mA.*

Table 7-1 *Model element values of the NE21938 at frequencies from 2 to 5 GHz.*

Element	Value	Units
L_b	1.70	nH
C_{be}	9.30	pF
R_{be}	4.1	Ω
R_{ce}	35	Ω
C_{ce}	1.70	pF
C_c	1.55	pF

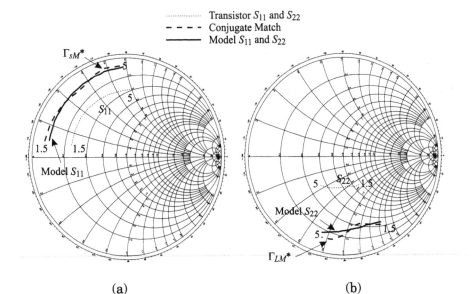

(a) (b)

Figure 7-3 *A plot of S_{11} (a) and S_{22} (b) of the NE21938 at a V_{ce} of 8 volts and a collector current of 8 mA, the conjugate match of the transistor, and the S_{11} and S_{22} of the unilateral transistor model from 1.5 to 5 GHz.*

A field effect transistor has a similar model. The main difference is in the gate-source junction, which is modeled as a series combination of an inductor, resistor, and capacitor. The reverse biased gate-source junction causes the capacitance, and the gate metalization causes the resistive part of the input circuit model. The bond wires create the series inductance on the gate. The full model of a field effect transistor is shown in Figure 7-4. We need to calculate the conjugate match of the FET and use these reflection coefficients to model the input and output. Figure 7-6 shows the input and output conjugate match of the NE76038. We have modeled the conjugate match of this transistor with the element values given in Table 7-2 for the model shown in Figure 7-5.

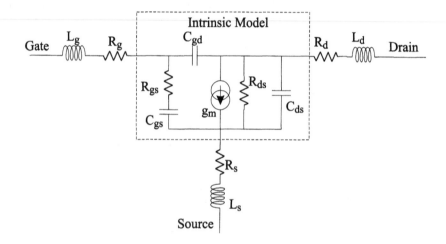

Figure 7-4 *The model of a FET.*

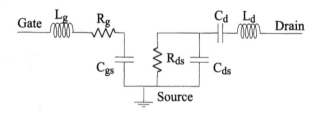

Figure 7-5 *A model for the conjugate match of the NE 67038 FET from 9 to 18 GHz.*

Table 7-2 *Model element values of the NE 76038 at frequencies from 9 to 18 GHz.*

Element	Value	Units
L_g	0.61	nH
R_g	5.5	Ω
C_{gs}	0.47	pF
R_{ds}	20	Ω
C_{ds}	0.48	pF
C_d	0.40	pF
L_d	0.44	nH

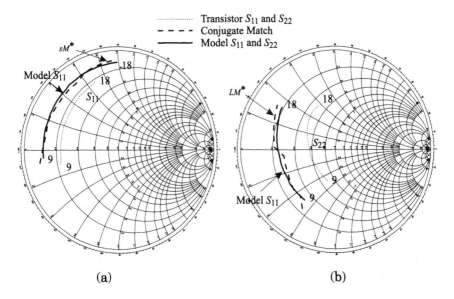

(a)   (b)

Figure 7-6 *The S_{11} (a) and S_{22} (b) of the NE67038 with $V_{ds} = 3V$ and a drain current of 10 mA, the conjugate matches of the FET, and the S_{11} (a) and S_{22} (b) of the unilateral transistor model of Figure 7-4(b) with the component values given in Table 7-2.*

A transistor input or output can be modeled with fewer components than those that are shown in Tables 7-1 and 7-2. Naturally, this will depend on the bandwidth of interest as well as the S-parameters of the transistor. The examples given above showed models that approximated the conjugate match over an octave bandwidth for the FET and almost two octaves for the bipolar transistor. Simpler models might be used over narrower bandwidths. Some broadband design techniques capitalize on the simplified model by assuming the transistor is a simple resistive load or a resistor-capacitor combination. When the transistor input or output can be approximated as a resistive load, a broadband quarter-wave transformer can be used to match the device.

Most transistors exhibit a drop in gain as the frequency increases. When the maximum available gain is plotted as a function of frequency, an approximate 6 dB/octave roll-off in gain can be observed. If the transistor was matched with the conjugate input and output match, the transducer gain of the amplifier would have the same 6 dB/octave gain roll-off with frequency. This is usually not desirable. We will show how to design amplifiers with gain that is fairly constant over the frequency band of interest.

7.3 Reactive Matching Gain-Bandwidth Limitations

This section introduces one method of determining the ability of a matching circuit to achieve a matched reflection coefficient over a given bandwidth. In previous chapters, we have been able to match any impedance perfectly at one frequency. In this section, we will look at the limitations of a circuit to simultaneously achieve gain and bandwidth. There is an upper limit to the ability of a reactive matching circuit to achieve a specified maximum gain, gain variation, ripple, or bandwidth. This limitation can be derived from Fano's integral [1], which is sometimes called the Fano bandwidth for certain matching circuits or circuit topologies. Synthesizing the input and output matching circuits to achieve a certain minimum gain, gain slope, or maximum gain ripple over a specified bandwidth is somewhat more difficult with broadband amplifiers.

We will begin by analyzing the problem of minimizing the reflection coefficient within a band of frequencies and maximizing it everywhere else. This problem is illustrated in Figure 7-7 where a circuit is desired to match a load to a source. Each circuit consists of a complex load, a two-port matching circuit, and a simple source impedance. As we have seen in Chapter 3, our ability to deliver power to a load or draw power from a source is directly related to the reflection coefficient. To maximize the power transfer between the source and load, the matching circuit should be a lossless reactive matching circuit that minimizes $|\Gamma_{in}|$. The techniques we have described in previous chapters show how to match a load to a single frequency. The value of the minimum reflection coefficient of $|\Gamma_{min}|$ that can be achieved over the frequency band from ω_1 to ω_2 is indicated by the solid line in Figure 7-8. The reflection coefficient is equal to 1 everywhere else. We also have to determine the minimum reflection coefficient for other bandwidths. A trade-off between gain and bandwidth can be expected, which will limit our ability to achieve a good match over a certain bandwidth.

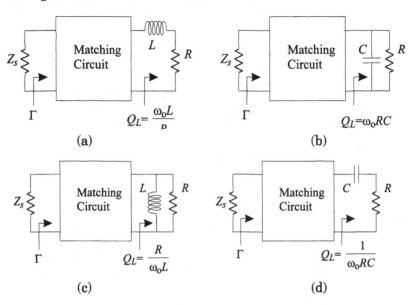

(a) $Q_L = \dfrac{\omega_o L}{D}$

(b) $Q_L = \omega_o R C$

(c) $Q_L = \dfrac{R}{\omega_o L}$

(d) $Q_L = \dfrac{1}{\omega_o R C}$

Figure 7-7 *Four simple load circuits with a two-port matching circuit to match the load to a source.*

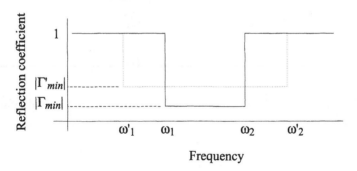

Figure 7-8 *Bandwidth and match trade-off using the same matching circuit topology.*

The fractional bandwidth is defined as

$$\omega' = \frac{\omega_2 - \omega_1}{\sqrt{\omega_1 \omega_2}} = \frac{\omega_2 - \omega_1}{\omega_o} \qquad 7.3$$

The input reflection coefficient will be a function of frequency $\Gamma(\omega)$. Fano's integral describes the limit on the reflection coefficient for the circuit types in Figure 7-7 (a) and (b) as a function of load circuit quality factor Q_L.

$$\int_0^\infty \ln\left|\frac{1}{\Gamma}\right| d\omega \le \frac{\pi\omega_o}{Q_L} \qquad 7.4$$

The area under the curve $\ln|1/\Gamma|$ cannot be greater than $\pi\omega_o/Q_L$ regardless of the type of lossless matching circuit used. The circuit type shown in Figure 7-7 (c) and (d) must satisfy the Fano integral.

$$\int_0^\infty \frac{1}{\omega^2} \ln\left|\frac{1}{\Gamma}\right| d\omega \le \frac{Q_L}{\pi\omega_o} \qquad 7.5$$

If we could achieve any reflection coefficient desired, the most efficient use of the area under the curve of Equation 7.4 or 7.5 would be a constant minimum Γ between the two band limits, ω_1 and ω_2, followed by a reflection coefficient equal to 1 everywhere else, which is shown in Figure 7-8. The minimum reflection coefficient in Equation 7.4 is related to load Q and fraction bandwidth ω' by

$$|\Gamma|_{\min} = \exp\left(-\frac{\pi\omega'}{Q_L}\right) \qquad 7.6$$

from which we determine the maximum gain.

$$G_{\max} = 1 - |\Gamma|_{\min}^2 = 1 - \exp\left(-\frac{2\pi\omega'}{Q}\right) \qquad 7.7$$

Ku and Peterson [2] have derived the optimum gain-bandwidth limitation

$$|\Gamma(\omega)|^2 = 1 - K_n \left(\frac{\omega}{\omega_2}\right)^\alpha \qquad 7.8$$

where K_n is the reflection coefficient function of the matching circuit at the upper frequency ω_2 ($1 \ge K_n$) and

$$\alpha = \frac{x}{10 * \log(2)} \qquad 7.9$$

where x is the desired gain slope in dB/octave. The gain of most transistors decreases by approximately 6 dB/octave. To compensate for this gain roll-off, we make $\alpha \cong 2$.

. The gain response in Figure 7-8 cannot be achieved by real matching networks. We can approximate the response by using a Chebyshev polynomial

$$G(\omega^2) = \frac{K_n}{1 + \varepsilon_n^2 T_n^2(\omega_b)} \qquad 7.10$$

where T_n is the nth order Chebyshev polynomial, ε_n is the ripple factor

$$\varepsilon_n^2 = 10^{\frac{ripple}{10}} - 1 \qquad 7.11$$

and ω_b is the bandpass-to-lowpass frequency transformation.

$$\omega_b = \frac{\sqrt{\omega_1 \omega_2}}{\omega_2 - \omega_1} \left(\frac{\omega}{\sqrt{\omega_1 \omega_2}} - \frac{\sqrt{\omega_1 \omega_2}}{\omega} \right) \qquad 7.12$$

Example 7.1: Find the minimum reflection coefficient that can be achieved from 1.5 to 5 GHz when designing a matching circuit for a parallel combination of a 10 pF capacitor and a 75 ohm resistor. The center frequency is

$$\omega_o = \sqrt{4\pi^4 (1.5E9)(5E9)} = 17.21E9 \ \text{rad/s}$$

The fractional bandwidth is

$$\omega' = \frac{2\pi(5E9) - 2\pi(1.5E9)}{\sqrt{4\pi^2 (1.5E9)(5E9)}} = 1.278$$

and the Q of the load is

$$Q_L = (\omega_o)(75)(10E - 12) = 12.91$$

The minimum reflection coefficient that can be achieved over this band is

$$|\Gamma|_{min} = \exp\left(-\frac{\pi\omega'}{Q_L} \right) = 0.0432$$

and the reduction on gain of amplifier would be

$$G_{max} = 1 - |\Gamma|_{min}^2 = 1 - \exp\left(-\frac{2\pi\omega'}{Q} \right) = 0.998$$

Equations 7.10 to 7.12 show that the flatness of the reflection coefficient in the passband is inversely related to the bandwidth for a given order of the Chebyshev polynomial. We will see that the number of circuit elements is proportional to the order of the Chebyshev polynomial. To widen the bandwidth, a higher level of mismatch must be tolerated or more circuit elements are needed. Principles of broadband design use techniques developed for the design of filters. Matching circuits based on the Chebyshev equations achieve the widest possible bandwidth for the lowest passband loss that is possible with a circuit made of series and shunt circuit elements.

7.4   Broadband Impedance Transformers

On occasion, the reactive elements of the transistor model may have little effect either because the desired bandwidth is narrow or their effect is small at the frequencies of interest. The matching circuit can be implemented with a broadband impedance transformer. Impedance transformers, introduced in Chapter 5, are one quarter-wavelength long at the center frequency. When a broader band transformer is needed, many quarter-wave transformers are used to transform from one impedance to another in steps using Butterworth or Chebyshev polynomials.

The methods given in this section work well if the impedance mismatch between the load and source is less than 2:1. Outside these bounds, tables published in [4] provide useful information. As a practical matter, the limitation on impedance ratio is usually sufficient for most applications. When the real part of the transistor impedance is too low or too high, the reactive elements will affect the impedance significantly. This would run contrary to the assumption that the reactive elements have little or no effect on the input or output impedance of the transistor.

The bandwidth of a single quarter-wave transformer depends on the impedance ratio between source and load. Below is the reflection coefficient of a transformer with an impedance of Z_t and load Z_L

$$\Gamma = \frac{Z_{in} - Z_s}{Z_{in} + Z_s} = \frac{Z_t(Z_L - Z_s) + j\tan\left(\dfrac{\omega\pi}{\omega_o 2}\right)(Z_t^2 - Z_s Z_L)}{Z_t(Z_L + Z_s) + j\tan\left(\dfrac{\omega\pi}{\omega_o 2}\right)(Z_t^2 + Z_s Z_L)} \qquad 7.13$$

$$= \frac{|Z_L - Z_s|}{Z_L + Z_s + j2\tan\left(\dfrac{\omega\pi}{\omega_o 2}\right)\sqrt{Z_s Z_L}} \qquad 7.14$$

where ω_o is the center frequency of the transformer. For frequencies near a perfect match, the magnitude of the reflection coefficient can be approximated by

$$|\Gamma| = \frac{|R-1|}{2\sqrt{R}}\left|\cos\left(\frac{\omega\pi}{\omega_o 2}\right)\right|$$

7.15

where R is the ratio of load and source impedance, $Z_L = RZ_s$. The fractional bandwidth of the quarter-wave transformer can be found for a maximum reflection coefficient.

$$\omega' = 2 - \frac{4}{\pi}\cos^{-1}\left|\frac{2|\Gamma|\sqrt{R}}{(R-1)\sqrt{1-|\Gamma|}}\right|$$

7.16

The 3 dB bandwidth of the impedance transformer is the frequency span between the points where the reflection coefficient is less than or equal to 0.5. One-half of the power is reflected from the transformer at the band edges. For other bandwidths such as the 1 dB bandwidth other values of $|\Gamma|$, e.g., 0.1, are used. Once the maximum tolerable reflection coefficient is determined, we can conclude that the bandwidth is a function of the input and output impedance ratio R. Figure 7-9 shows a plot of the 3 dB bandwidth of a single quarter-wave transformer as a function of the impedance ratio R.

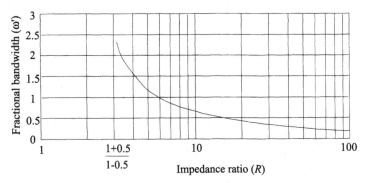

Figure 7-9 *The fractional bandwidth of a quarter-wave transmission line as a function of the input and output impedance ratio with $\Gamma = 0.5$.*

Given a desired bandwidth with lower and upper frequencies of f_1 and f_2, and a maximum allowable reflection coefficient, there is a minimum number of quarter-wave transformers that can achieve this reflection coefficient. However, other than a knowledge of the minimum number of transformers needed and an understanding that each one is a quarter-wavelength long, there is no unique solution for the characteristic impedance of the transformers. The common method of determining the impedances is by using what is called a *maximum flatness* or an *equal-ripple response*. The maximum flatness method finds the combination of impedances that gives the flattest possible reflection coefficient over the band. This is done using impedances based on a Butterworth polynomial. An equal-ripple transformer allows some fluctuation in the reflection coefficient over the band. Furthermore, all these fluctuations reach equal peaks across the band. The transformer impedances are based on coefficients of a Chebyshev polynomial, which provides the widest possible bandwidth given a specific number of elements and a maximum allowable ripple. The Butterworth and Chebyshev polynomial were derived mathematically to achieve the maximum flatness or equal-ripple function for the widest possible bandwidth. There are no other functions that approximate these conditions better than these two.

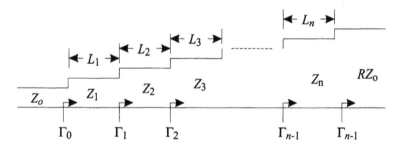

Figure 7-10 *A diagram of a multiple-section quarter-wave transformer using a series of transmission lines that are one-fourth of a wavelength long.*

Figure 7-10 contains a diagram of an impedance transformer constructed from a series of quarter-wave transformers to match the impedance Z_o to RZ_o. Equation 7.15 shows that the bandwidth of a single quarter-wave transmission line is inversely proportional to the impedance ratio. The bandwidth is wider when the input-to-output impedance ratio is smaller. If we use many sections, as shown in Figure 7-10, each step can be small, and we can provide a bandwidth wider than a single section of a quarter-wave transformer. It is necessary to know how many sections to use, the impedance, and the length of each section. The length of each transmission line is one-fourth of a wavelength long at the center frequency of the transformer.

$$L_i = \frac{\lambda_{ih}\lambda_{il}}{2(\lambda_{ih} + \lambda_{il})} \qquad 7.17$$

λ_{ih} and λ_{il} are the wavelengths at the high and low frequencies of the band in the transmission line with an impedance Z_i. If the wavelength does not depend on the impedance of the transmission line, all the lengths will be the same. When the wavelength is dependent on the characteristic impedance, as it is the case for a microstrip transmission line, the lengths of each quarter-wave section (L_i) will be different.

The number of sections is related to the bandwidth and the maximum acceptable loss within the passband. We obtain a wider bandwidth as we increase the order of the Butterworth polynomial. However, we also increase the number of impedance sections in the transformer. It is desirable to decrease the number of transmission lines in the matching circuit for the applications. The Butterworth polynomial has the special distinction of having the flattest possible response that can be achieved with a certain number of sections. We want to design a matching circuit with a reflection coefficient in the form of the Butterworth polynomial.

$$\Gamma = A\left(1 + e^{-2j\frac{\omega\pi}{2\omega_o}}\right)^N \qquad 7.18$$

The scale factor A is chosen to ensure that the reflection coefficient is correct when the frequency is equal to zero

$$\Gamma = \frac{Z_L - Z_s}{Z_L + Z_s}\bigg|_{\omega=0} \qquad 7.19$$

so that

$$A = 2^{-N}\frac{Z_L - Z_s}{Z_L + Z_s} \qquad 7.20$$

We can use the binomial expansion on the reflection coefficient

$$\Gamma = 2^{-N}\frac{Z - Z}{Z + Z}\sum_{n=0}^{N}C_n^N e^{-j2n\frac{\omega\pi}{2\omega_o}} \qquad 7.21$$

where C_n^N are binomial coefficients given by

$$C_n^N = \frac{N(N-1)(N-2)\cdots(N-n+1)}{n!} = \frac{N!}{(N-n)!n!} \qquad 7.22$$

The magnitude of each reflection coefficient at the junctions of the quarter-wave transmission line is given by

$$|\Gamma_n| = 2^{-N} \left| \frac{Z_L - Z_s}{Z_L + Z_s} \right| C_n^N \qquad 7.23$$

When the reflection coefficients (Γ_n) are small, the characteristic impedance of each section is approximated by

$$\ln \frac{Z_{n+1}}{Z_n} \approx 2 \frac{Z_{n+1} - Z_n}{Z_{n+1} + Z_n} = 2|\Gamma_n| \qquad 7.24$$

$$= 2^{-N} C_n^N \ln(R) \qquad 7.25$$

which should only be used in the range of $0.5 > R > 2.0$ due to the approximation in Equation 7.24.

$$\ln(R) \approx 2 \frac{Z_L - Z_s}{Z_L + Z_s} \qquad 7.26$$

When the impedance ratio is outside these bounds, other sources [4] that offer more exact formulas or tables of coefficients for the step impedances should be consulted. The minimum number of sections can be found for a specified bandwidth (ω') and the largest tolerable reflection coefficient ($|\Gamma_{max}|$). The absolute value of Equation 7.18 is combined with Equation 7.20 to relate ω', $|\Gamma|$, R, and N.

$$\omega' = 2 - \frac{4}{\pi} \cos^{-1} \left| \frac{2|\Gamma_{max}|}{\ln(R)} \right|^{\frac{1}{N}} \qquad 7.27$$

The Butterworth response will be very flat in the center of the passband and rapidly become lossy at the band edges.

The Chebyshev polynomial gives a passband response that has some variation, or ripple, in loss across the passband. Sometimes gain ripple is permitted as long as it is less than a certain value and the Chebyshev response exhibits this equal-ripple loss characteristic. The Butterworth circuit will have the entire gain variation at the band edges, whereas the Chebyshev circuit will have ripple across the entire band. However, the Chebyshev response will have a larger bandwidth than the Butterworth circuit for the same amount of ripple in the passband. The Chebyshev polynomial of degree one to n is given recursively by

$$T_1(x) = x \qquad\qquad 7.28$$

$$T_2(x) = 2x^2 - 1 \qquad\qquad 7.29$$

$$T_3(x) = 8x^4 - 8x^2 + 1 \qquad\qquad 7.30$$

$$T_n(x) = 2xT_{n-1}(x) - T_{n-2} \qquad\qquad 7.31$$

where the higher-order polynomials are found recursively from the last two lower-order polynomials.

The reflection coefficient is forced to equal a Chebyshev polynomial in the passband

$$\Gamma = Ae^{-2j\frac{\omega\pi}{2\omega_o}} T_N\left(\sec(\theta_m)\cos\left(\frac{\omega\pi}{2\omega_o}\right)\right) \qquad\qquad 7.32$$

where

$$\theta_m = \frac{\pi}{2} - \frac{\pi}{4}\omega , \qquad\qquad 7.33$$

and A is a constant. When $\omega = 0$, the reflection coefficient is unaffected by the transformer

$$\Gamma = \frac{Z_L - Z_s}{Z_L + Z_s} = AT_N\left(\sec\theta_m\right) \qquad\qquad 7.34$$

which gives us an expression for A.

$$A = \frac{Z_L - Z_s}{(Z_L + Z_s)T_N\left(\sec\theta_m\right)} \qquad\qquad 7.35$$

The maximum passband ripple caused by a nonzero reflection coefficient and bandwidth are related by

$$\left|\Gamma_{ripple}\right| = A = \frac{Z_L - Z_s}{(Z_L + Z_s)T_N\left(\sec\theta_m\right)} \qquad\qquad 7.36$$

The passband, maximum ripple, and order of the polynomial are all related by

$$T_N\left(\sec\theta_m\right) \geq \frac{Z_L - Z_0}{Z_L + Z_0}\left|\Gamma_{ripple}\right|^{-1} \qquad\qquad 7.37$$

where N is the smallest integer giving a fractional bandwidth of ω' or greater.

The lengths of the quarter-wave transmission lines are the same as given by Equation 7.17. The impedances are found by expanding the Chebyshev polynomial in Equation 7.35

$$\Gamma = 2e^{-2j\frac{\omega\pi}{2\omega_o}}\left[|\Gamma_1|\cos\left(N\frac{\omega\pi}{2\omega_o}\right) + |\Gamma_2|\cos\left((N-2)N\frac{\omega\pi}{2\omega_o}\right)+\cdots\right.$$

$$\left. +|\Gamma_n|\cos\left((N-2n)\frac{\omega\pi}{2\omega_o}\right)\right]$$

7.38

where

$$\Gamma_n = \frac{Z_{n+1} - Z_n}{Z_{n+1} + Z_n}$$

7.39

Example 7.2: Find the maximally flat impedance transformer to match a 35 ohm load to a 50 ohm source from 2 to 5 GHz with a reflection coefficient of no more than 0.1. First, we calculate R, ω_o and ω'.

$$R = \frac{Z_L}{Z_s} = \frac{35}{50} = 0.70$$

$$\omega_o = \sqrt{4\pi^2(2E9)(5E9)} = 1.987E9$$

$$\omega' = \frac{2\pi(5E9 - 2E9)}{\omega_o} = 0.9487$$

Next, we need to find the number of sections from Equation 7.27. We assume that any loss in the matching circuit is due to the reflection coefficient. A reflection loss of 0.1 dB corresponds to a minimum circuit gain of

$$G_{min} = 10^{-0.02} = 0.9772$$

and a maximum reflection coefficient of

$$|\Gamma_{max}| = \sqrt{1-G_{min}} = \sqrt{1-0.9772} = 0.1509$$

$$0.9487 = 2 - \frac{4}{\pi} \cos^{-1} \left| \frac{2(0.1)}{\ln(0.7)} \right|^{\frac{1}{N}}$$

$$\cos\left(\frac{\pi}{4}(2 - 0.9487)\right) = 0.6780 = \left| \frac{2(0.1)}{\ln(0.7)} \right|^{\frac{1}{N}} = |-0.5607|^{\frac{1}{N}}$$

We take the logarithm of both sides and solve for N.

$$\ln(0.6780) = \frac{1}{N} \ln(0.5607)$$

$$N = \frac{\ln(0.5607)}{\ln(0.6780)} = 1.488$$

We need at least two transmission line sections to achieve a reflection coefficient of 0.1 or lower at the band edges. We then calculate A.

$$A = 2^{-2} \frac{35 - 50}{35 + 50} = -0.0441$$

The binomial coefficients are determined and we can obtain Γ_0 and Γ_1.

$$C_0^2 = \frac{2!}{2!} = 1$$

$$C_1^2 = \frac{2!}{1!} = 2$$

$$|\Gamma_0| = 2^{-2} \left| \frac{35 - 50}{35 + 50} \right| 1 = 0.0441$$

$$|\Gamma_1| = 2^{-2} \left| \frac{35 - 50}{35 + 50} \right| 2 = 0.0881$$

The first transmission line will create a reflection coefficient of 0.0441 when connected to the 50 ohm source.

$$Z_1 = 50 \frac{1 - 0.0441}{1 + 0.0441} = 45.78\Omega$$

The second transmission line should create a reflection coefficient of 0.0882 when connected to the 45.78 ohm transmission line.

$$Z_2 = 45.78 \frac{1 - 0.0882}{1 + 0.0882} = 38.36\Omega$$

We can calculate the reflection coefficient between the 35 ohm load and the 38.36 ohm transmission line to check our math.

$$\Gamma_2 = \frac{38.36 - 35}{38.36 + 35} = 0.0440$$

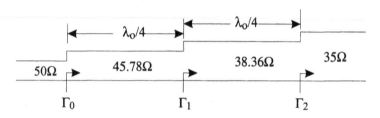

Figure 7-11 *The impedance-matching circuit for Example 7.2.*

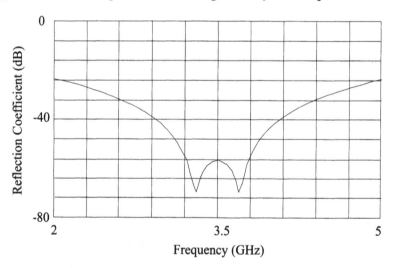

Figure 7-12 *The frequency response of the circuit in Figure 7-10, as simulated by Eagleware's =SuperStar=.*

Tapered transformers, which can have a smaller overall size at the same bandwidth of a stepped quarter-wave transformer, can be used. They are synthesized as though there is an infinite number of transmission lines, each with an infinitely short length.

7.5. Reactive Matching

Reactive matching describes matching circuits that have no resistive elements. These networks ideally have no loss and are constructed from capacitors, inductors, and transmission lines. In reality, there will be some loss, but we will treat these circuits as though they were lossless. The topology of the matching circuit will be carefully chosen so that the transistor model capacitance or inductance is absorbed into the matching circuits. The input and output transistor circuit models will be used to choose the circuit topology. There are three useful topologies that can be used depending on the circuit model of the transistor or amplifying device: bandpass, lowpass and highpass. Figure 7-13 shows the circuit prototypes for these three matching networks. The network is designed to match the source or load resistance, usually 50 ohms, to the equivalent resistance of the device. The reactive parts of the transistor input or output impedance model are used to replace some or all of the circuit elements closest to the device.

Consider the input of a FET modeled in Figure 7-5. We can rearrange the resistor and capacitor, producing a series combination that is compatible with the bandpass topology shown in Figure 7-13(a). A prototype bandpass filter circuit is designed using a Butterworth or Chebyshev filter response that has a series capacitor and inductor as the circuit elements near the transistor. The input capacitance of the FET can be absorbed in the series capacitor of the matching circuit.

For the output matching circuit, we have assumed that the series inductance in the transistor model has little or no effect on the performance of the circuit at these frequencies. Using a highpass topology matching circuit offers an excellent opportunity to compensate for the 6 dB/octave gain roll off of the device. The result is a lumped-element matching circuit that may match the transistor over the designed band, as illustrated in Figure 7-14. The output model of the transistor matches the first element of a lowpass prototype filter circuit. The drain capacitance is absorbed into the first shunt capacitor of the prototype filter circuit.

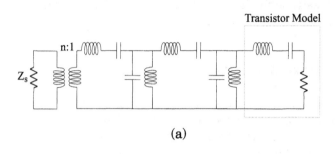

(a)

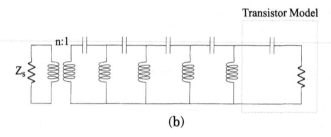

(b)

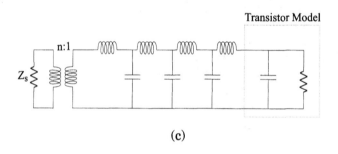

(c)

Figure 7-13 *A bandpass (a), highpass (b), and lowpass (c) matching circuit topology.*

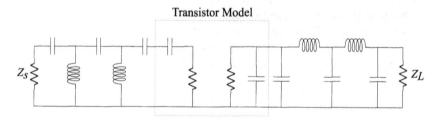

Figure 7-14 *A broadband input and output matching circuit for a FET with an input circuit model consisting of a series capacitor. The resistor and the output FET are modeled as a parallel resistor and capacitor.*

The topology of the matching circuit will depend on the transistor model. A model that approximates the frequency response of the transistor is desirable. Sometimes the reactive elements cannot be subtracted from the circuit prototype because it would result in a negative capacitance or inductance. Instead, another circuit can be designed using a higher-order polynomial. The design procedure is described by Fano [1] and Levy [3] and contains four steps:

1. Design a transistor circuit model that approximates the frequency response of the transistor and determine the minimum number of filter elements that will give the desired bandwidth and reflection coefficient.

2. Synthesize a lowpass matching circuit just as you would synthesize a filter between the transistor and the source, or load. This circuit will usually contain an ideal transformer.

3. Perform any lowpass-to-bandpass or lowpass-to-highpass transformation.

4. Replace the ideal transformer with an equivalent T or p network as described by Levy [3]. If a negative inductance or capacitance results, increase the number for filter elements and repeat steps 2 through 4.

Lumped elements are used for lower frequency and monolithic circuits. High-frequency microwave and millimeter-wave microstrip circuits are realized by converting the lumped-element capacitor and inductors to transmission lines. Hence, the circuits above are called circuit prototypes. Series inductors can be approximated by a series transmission line using the highest possible impedance (Z_{oh}) with an electrical length of

$$\phi = \frac{180}{\pi} \arcsin\left(\frac{\omega_2 L_s}{Z_{oh}}\right) \qquad 7.40$$

where ω_2 is the high-frequency band edge. Shorted or open transmission lines can approximate shunt elements. A shorted transmission stub can be used to approximate a shunt inductor by using the formula

$$Z_{sc} = iZ_o \tan\phi \qquad 7.41$$

where ϕ is the electrical length of the transmission line at the center of the band and is less than a quarter wavelength (usually 45°). An open transmission line is used to approximate a shunt capacitor using

$$Z_{oe} = \frac{-iZ_o}{\tan(\phi)} \qquad\qquad 7.42$$

where ϕ is the electrical length of the open stub at midband and usually equals 45°. There is no sufficient approximation to a series capacitor in transmission line form.

Two shunt transmission lines can approximate shunt resonant circuits: one open and one shorted. Or, both shunt stubs can be open transmission lines, one a quarter-wavelength longer at the center frequency than the other. Figure 7-15(a) shows a combination of two shunt transmission lines. At resonance, the shorter transmission line is designed to be 45 electrical degrees long with an impedance of $-iZ_o$; the long stub has an impedance of $+iZ_o$. The two impedances cancel. At lower frequencies, the capacitance of the shorter stub becomes greater than the inductance of the longer one. The capacitance of the shorted stub is no longer canceled by the inductance of the longer stub, and the circuit acts as a shunt capacitance. At higher frequencies, the capacitance of the shorted stub decreases as the inductance of the longer stub increases. The circuit becomes inductive. The impedance of the double-stub circuit is shown in Figure 7-15(b) where the impedance of the circuit is shown plotted on a Smith chart.

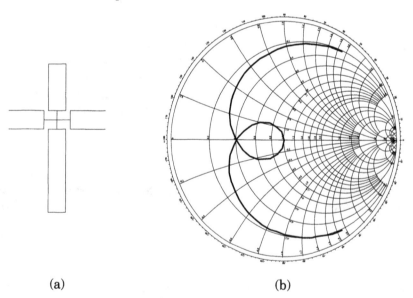

(a)   (b)

Figure 7-15 *A double-stub resonant tuning structure and the impedance with the impedance plotted on a Smith chart.*

7.6   Resistive Matching Circuits

The last two circuit synthesis techniques use lossless transmission lines or components to match the transistor to the source and load. This section describes a technique that uses lossy matching circuits to match the transistor and compensate for the gain roll-off with frequency. Consider a matching network consisting of a shunt transmission line with a grounded resistor at the end, as shown in Figure 7-16(a). At very low frequencies, the transmission line will have little or no effect on the impedance of the matching element. This circuit will resemble a shunt resistor. As the frequency of excitation increases, the transmission line will begin to rotate the resistance around the Smith chart, changing the shunt impedance with frequency, as shown in Figure 7-16. The proper match may be possible over a range of frequencies with the right combination of transmission line impedance, length and termination resistor.

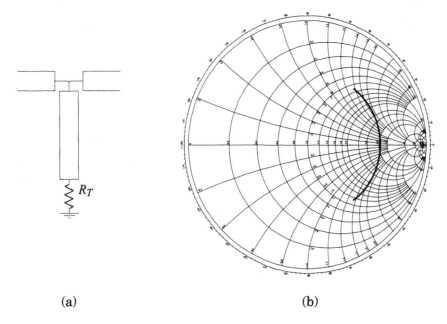

(a)   (b)

Figure 7-16 *A resistively loaded transmission line.*

Example 7.5: An amplifier was designed using a M/A-Com MA4F004 medium power FET using resistive matching. Figure 7-17 shows the input and output matching circuits.

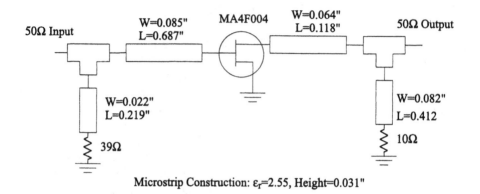

Figure 7-17 *The microstrip circuit of a broadband amplifier using resistive matching on the input.*

The circuit was simulated using =SuperStar=. The gain and the return loss are shown in Figure 7-18.

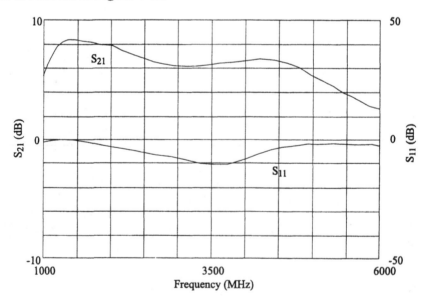

Figure 7-18 *The gain and return loss of the broadband amplifier in Figure 7-17.*

The measured frequency response from 1 to 6 GHz is shown in Figure 7-19.

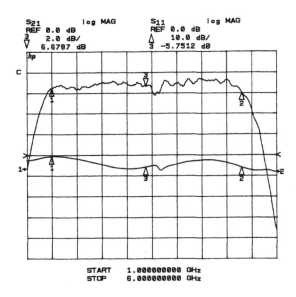

Figure 7-19 *The measured frequency response of the circuit.*

7.7   Active Matching

Active matching uses other transistors in the matching circuit and is well-suited for monolithic circuits. Space on a monolithic circuit is very valuable, and it may not be practical to build the matching circuits from transmission lines, capacitors, or inductors. Transistors are small and provide an efficient way to achieve a broadband match for microwave monolithic circuits. Figure 7-20 shows a FET that is matched on the input and output by two other FETs. The FET in the center is the one we will match to a 50 ohm source and a 50 ohm load. A common-gate FET is used on the input, and a common-drain FET is used on the output.

The FET has a high input impedance compared to a 50 ohm source impedance. A common-gate FET has a low input impedance and a high output impedance. A transistor in a common gate configuration can be used to transform the relatively low source impedance to a higher one to match the FET T_2 shown in Figure 7-20. This common-gate FET would be manufactured with dimensions specifically designed for an input matching circuit.

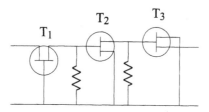

Figure 7-20 *Active-matching networks on the input and output of a FET.*

The output impedance of a transistor is usually very low. Another transistor can be used on the output to transform the low impedance of T_2 to a higher impedance closer to 50 ohms. A FET in a common-drain configuration is used as the output matching circuit. The dimensions of this output matching FET are designed strictly for matching the output of the common-source FET.

Most microwave computer-aided design programs can easily calculate the common-gate or common-drain S-parameters from the common-source S-parameters that are published. Equations for the transformation of two-port S-parameters to the three-port S-parameters are given in Equations 4.42 through 4.50. The transistor two-port is transformed to a three-port with a series of equations. The three ports of a FET are the gate, source, and drain where none of the transistor terminals are connected to ground. The two-port S-parameters for the common-gate or common-drain configuration are obtained by connecting a short with $\Gamma = -1$ to the gate or drain respectively. The new two-port S-parameters then are found using methods of calculating S-parameters with one port terminated, as described in Chapter 3.

7.8 Balanced and Distributed Amplifiers

This section briefly describes balanced and distributed amplifier topology. Balanced amplifiers refer to the use of two power splitters and two amplifiers to build a single amplifier stage, as shown in Figure 7-21. If the two amplifiers are identical, the reflected powers can be forced to cancel each other out in the power splitter instead of going back through the input port. The input reflection coefficient of the balanced amplifier can be improved compared to an individual amplifiers. The same happens at the output, giving the output a good match as well. The balanced amplifier can achieve a good match over the effective bandwidth of the power splitter.

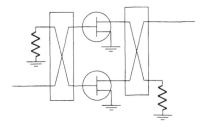

Figure 7-21 *The schematic of a balanced amplifier.*

Distributed amplifiers can have an ultrawide bandwidth. A distributed amplifier topology is based on a transmission line feed network. Figure 7-22 shows a simplified schematic of a distributed amplifier. The gates are driven by a tapped transmission line that usually has fairly high impedance and is less than one-fourth of a wavelength long at the highest frequency between taps. Each transmission line acts as a series inductance. The gate of each transistor acts as a shunt capacitor. The series inductances and shunt capacitances form a transmission line similar to the one shown in Figure 3-3. The signals of many frequencies can propagate along this transmission structure, driving the inputs of the transistors. The outputs of the transistors form the shunt capacitors of another transmission line structure. Again, series high-impedance transmission lines act as inductors. The impedance of the input and output of the transmission lines can be approximated using Equation 3.25. We want the input transmission line structure formed by the transistor gates and series high-impedance line to approximate the input impedance Z_i. If the equivalent input capacitance of the transistors is C_g, the series inductance of a Z_i impedance transmission can be obtained from Equation 3.25.

$$L_s = C_g Z_i^2 \qquad\qquad 7.43$$

The length of the transmission line is found by

$$\phi = \frac{180}{\pi} \arcsin\left(\frac{\omega^2 C_g Z_1^2}{Z_{oh}} \right) \qquad\qquad 7.44$$

where Z_{oh} is the impedance of the series transmission line and is usually chosen as the highest practical impedance that can be fabricated.

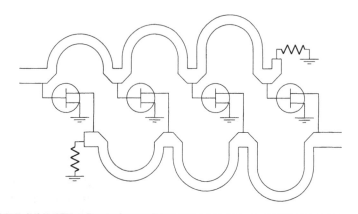

Figure 7-22 *The schematic of a distributed amplifier and a plot of the gain of the amplifier from 1 to 20 GHz.*

7.9   Summary

In this chapter, we introduced various broadband amplifier design techniques. We can commonly achieve bandwidths of 5% using the narrowband techniques. Wideband amplifiers have bandwidths of 50-100%, and special methods are usually required to design these circuits. Ultrawideband amplifiers have bandwidths of greater than 100% and will always require special design methods.

The key to the methods used in this chapter is the design of a good transistor model. We used the equivalent model instead of a collection of S-parameters to leverage low-frequency circuit synthesis techniques. The performance of the amplifier will be highly dependent upon the accuracy of the equivalent model for the transistor. Useful models are those based on the transistor two-port S-parameters or the conjugate match. Most manufacturers publish the model for their transistors. These models can be used either as they are given or be simplified. It is usually more efficient to use the published model than to develop a model based on the conjugate match. If a transistor has no model, many published techniques for the development of transistor models can be utilized. The need to research these methods and the selection of one or more techniques will depend on how often transistors need to be modeled. Most model development methods will take time before the engineer becomes proficient.

The bandwidth of a matching circuit is mathematically related to the mismatch of the transistor and the complexity of the matching circuit. This imposes a fundamental limit on the gain-bandwidth product of a particular circuit topology. Amplifier gain depends on the amount of gain that can be obtained from the transistor, the maximum stable gain, or maximum available gain. A mismatch on the input and output reduces the gain. The product derived from the quality of the match and the bandwidth is constant for a particular circuit topology. To achieve a greater bandwidth or better match, more circuit elements have to be used.

Section 7.4 described the design of a broadband quarter-wave impedance transformer. These circuits are useful when a simple resistive load can approximate the input or output of the transistor. Equations were presented that are useful for designing impedance transformers for source-to-load ratios between 0.5 and 2.0.

Section 7.5 showed the use of filter design techniques in the design of broadband amplifier matching circuits. Many filter design texts provide excellent information on designing broadband amplifiers [5, 6]. Filter design methods include the capability for designing filters with different source and load impedances. The reactive components of the transistor model are used as a filter prototype and are absorbed in the like elements of the matching circuit.

Section 7.6 introduced resistive matching as a simple and effective method of broadband design. The matching circuits are designed with lossy elements to lower the Q of the reactive elements in the matching circuit. The limitation imposed by the Fano bandwidth criteria depend on the circuit Q. Resistors in the matching circuits lower the circuit Q, allowing for a wider bandwidth. These circuits should not be used in high-power amplifiers or low noise amplifiers. Lossy elements in high power amplifiers lower the efficiency and could generate excessive heat. Lossy elements also add noise to the signal and should not be used in low noise amplifiers.

We introduced active matching in Section 7.7. Monolithic microwave circuit designs should consume as little chip space as possible. Active matching is used to save space. It is relatively easy to build common-gate or common-drain transistor stages on a monolithic substrate. A common-gate transistor has a high input impedance and a low output impedance. Placing a common-gate transistor in the input matching circuit makes it less difficult to match the higher 50 ohm source to the high input impedance of the amplifier transistor. The same holds for the output matching circuit where the relatively low output impedance of the gain transistor has to be matched to the 50 ohm load.

In Section 7.8 two broadband techniques were discussed: balanced and distributed amplifiers. Balanced amplifiers utilize power dividers at the input and combiners at the output of a transistor amplifier pair. Reflections from the transistors are tolerated because the reflected power is terminated in the power splitter or combiner. The distributed amplifier uses the capacitance of the transistors as part of a transmission line structure on the input and output of the amplifier. The power propagates along the transmission line feeding power to, or collecting power from, each transistor in the amplifier.

7.10   Problems

7.1   Find the minimum reflection coefficient that can be achieved over a 2 to 3 GHz bandwidth when designing a circuit to match a load to 50 ohms (see Figure 7-6), with a series combination of a 6 ohm resistor and a 1.4 pF capacitor.

7.2   What is the Fano bandwidth limit, centered at 14 GHz, to achieve a reflection coefficient of 0.1 with the output circuit of the FET model in Figure 7-5 and component values of R_{ds} = 20 ohms, C_{ds} = 0.48 pF and C_d = 0?

7.3   What is the maximum gain that can be achieved with the transistor modeled in Figure 7-23 over an octave bandwidth centered at 8 GHz? This special transistor has a maximum narrowband gain of 20 that is constant over frequency.

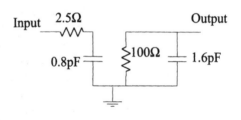

Figure 7-23 *A model transistor for Problem 7.3.*

7.4   Design a maximally flat (Butterworth) impedance transformer to match a 75 ohm load to a 40 ohm source that operates from 2 to 12 GHz.

7.5   Design an equal ripple (Chebyshev) impedance transformer to match a 75 ohm load to a 50 ohm source that operates from 1 to 2 GHz with no more than 0.5 dB of ripple.

7.11   References

1.   Fano, R. M. "Theoretical Limitations on Broad-Band Matching of Arbitrary Impedances." *J. Franklin Institute* 249 (1950): 57.

2.   Ku, W. H. and W. C. Peterson. "Optimum Gain-Bandwidth Limitations of Transistor Amplifiers as Reactively Constrained Active Two-Port Networks." *IEEE Transaction on Circuit Systems* 22, no. 6 (1975): 523.

3. Levy, R. "Explicit Formulas for Chebyshev Impedance-Matching Networks, Filters and Interstages." *Proceeds of the IEE* 111, no. 6 (1964): 1099.

4. Mathaei, George L., Leo Young, and E. M. T. Jones. *Microwave Filters, Impedance-Matching Networks, and Coupling Structures*. New York: McGraw-Hill, 1964.

5. Temes, G. C. and J. W. LaPatra. *Circuit Synthesis and Design*. New York: McGraw-Hill, 1977.

6. Williams, Arthur B. and Fred J. Taylor. *Electronic Filter Design Handbook*. New York: McGraw-Hill, 1995.

Introduction to Noise

8.1   Introduction

Noise analysis of microwave circuits and systems has been a research issue for many decades. A circuit's noise performance is given in terms of noise figure or noise temperature. The signal energy exiting a generator or antenna is amplified or attenuated in passing from the input to the output of a two-port network. Along this signal path, thermal energy in the components add noise to the signal. In some applications, e. g., in a radar receiver, it is necessary to anticipate the amount of noise that circuits add to the signal path. When receiving a radar pulse, the receiver will also add some noise to the signal. This added noise masks the pulses, which will it make more difficult to detect them. When the ratio of pulse power to noise becomes too low, the pulse and the target cannot be detected. Thus, a receiver that adds the least amount of noise possible is desirable. There is a similar requirement for a microwave radio. A microwave signal is modulated with data and transmitted from an antenna on top of a mountain. A receiver on a neighboring mountain ridge receives this signal and extracts the data. When the signal becomes overwhelmed by noise, however, the data stream can no longer be reconstructed.

Section 8.2 introduces a model for the generation of random noise. We will study *thermal noise*, which has constant power at any frequency of interest to us. Thermal noise can be modeled as voltage and current sources that output a wide spectrum of signals with a well-defined power spectrum. A noise model will be developed for two-port networks. We will discuss how the noise power is affected by the input reflection coefficient.

Section 8.3 discusses the use of noise sources to noise power waves. The ratio of the forward- and reverse-traveling power waves is used in the definition of S-parameters. Since S-parameters are the primary method used to characterize microwave circuits, noise waves have been developed as a natural method to model noise. This model represents powerwave generators that inject noise into a network as traveling waves.

Section 8.4 introduces noise figure as a simplified method of characterizing the noise performance of a network. In order to describe the noise in a network, we need to characterize the noise sources as well as the network parameters. In most cases, all we need to know is how the signal-to-noise ratio changes by passing through a circuit. Noise figure effectively removes any gain from the calculation of the amount of noise added by the amplifier. Noise figure is referenced to the input of the circuit and is only affected by the input reflection coefficient.

Finally, Section 8.5 shows noise figure measurement techniques. We must be able to measure noise figure to characterize transistors and circuits. Noise is measured with sensitive power meters and a well-characterized source of noise. The most common method of measuring noise figure uses two noise sources with a known excess noise ratio (ENR), which makes it possible to measure the noise figure and gain of an amplifier simultaneously.

8.2   Noise Model

Thermal noise in microwave circuits and microwave amplifiers manifests itself by generating random voltage and current [10, 11]. The open-circuit noise voltage or short-circuit noise current has a time average value of zero. However, the noise voltage and current have a finite mean-squared value that depends on the thermal temperature of the source. A resistance R in thermal equilibrium at a temperature T will have an open-circuit mean square noise voltage in a bandwidth B of

$$\left\langle \left| e_n \right|^2 \right\rangle = 4kTRB \qquad\qquad 8.1$$

where the brackets <> indicate the mean value of the quantity inside over time and k is Boltzman's constant (1.374E–23 J/K). The quantity $4kTR$ is the spectral power density of the noise generated by the source. Alternatively, the short-circuit mean square noise current is equal to

$$\left\langle \left| i_n \right|^2 \right\rangle = 4kTGB \qquad\qquad 8.2$$

where G is the conductance of the thermal source. These equations give the mean square voltage and current generated in a resistance or conductance at a finite thermal temperature. The available noise power that a noise source at temperature T can deliver to a matched load is given by

$$P_n = \frac{\langle |e_n|^2 \rangle}{4R} = \frac{R}{4} \langle |i_n|^2 \rangle = kTB \qquad 8.3$$

These relations indicate that heat or the thermal agitation of molecules in a resistance generate an electromotive force, which, in turn, generates random voltage and current fluctuations. These random signals have a zero mean and are spread over all frequencies that interest us. The equations are applicable up to about 1000 GHz at temperatures less than a few million degrees kelvin.

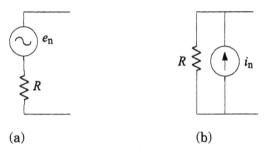

(a)   (b)

Figure 8-1 *The open-circuit noise voltage (a) and the short-circuit current (b) of a resistor in thermal equilibrium.*

We will focus on the study of noisy two-ports that can either be active or passive and assume that the noise power generated in a network is independent of any external excitation. We can model a noisy two-port as a combination of a perfect noiseless two-port with external noise sources [9]. The noisy two-port is represented as a noiseless two-port and external current and/or voltage generators, as shown in Figure 8-2. The linear transfer equations for the two-port in Figure 8-2(a) are

$$I_1 = Y_{11}V_1 + Y_{21}V_2 + i_{n1} \qquad 8.4$$

$$I_2 = Y_{21}V_1 + Y_{22}V_2 + i_{n2} \qquad 8.5$$

If we prefer to use Z-parameters, the noise is easily expressed in terms of equivalent voltage sources. For the circuit in Figure 8-2(b), the equations for the two-port are

$$V_1 = Z_{11}I_1 + Z_{21}I_2 + e_{n1}$$ 8.6

$$V_2 = Z_{21}I_1 + Z_{22}I_2 + e_{n2}$$ 8.7

In Equations 8.4 through 8.7, the Y- and Z-parameters describe the two-port's linear voltage and current response. All noise coming from the network's thermal temperature is modeled by the noise sources in Figures 8-2(a) and 8-2(b).

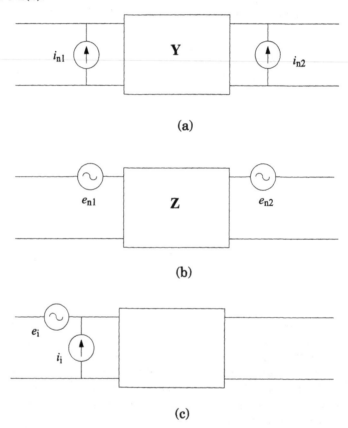

(a)

(b)

(c)

Figure 8-2 *Noise models using equivalent noise-current generators in (a), noise-voltage generators in (b) and input noise sources (c).*

Usually, it is more convenient to place noise sources on the input [Figure 8-2(c)]. We need both noise current and voltage sources because noise still is generated by the two-port when the input is either short-circuited or open-circuited. With the introduction of noise sources i_i and e_i, Equations 8.4 and 8.5 become

$$I_1 = Y_{11}(V_1 - e_i) + Y_{21}V_2 + i_i \qquad 8.8$$

$$I_2 + Y_{21}(V_1 - e_i) + Y_{22}V_2 \qquad 8.9$$

Or, in terms of Z-parameters, the input noise sources are included as

$$V_1 = Z_{11}(I_1 - i_i) + Z_{21}I_2 + e_i \qquad 8.10$$

$$V_2 = Z_{21}(I_1 - i_i) + Z_{22}I_2 \qquad 8.11$$

Noise generated in the two-port is modeled by two noise sources each with a zero time average but with a finite power. The noise power supplied by these sources will add to the noise power of the signal source. In general, the two noise sources are not entirely independent. Let us use the input reference noise model shown in Figure 8-2(c). One internal source of noise can create both current and voltage effects at the network's input and output terminals. The random voltages generated by e_i will have some effect on the amount of current generated by i_i. In addition, the reverse is true about the effect of current i_i on the voltage source e_i. These sources are somewhat correlated. The noise voltage source e_i can be separated into one part that is uncorrelated with the noise current source i_i, and one part that is correlated with i_i. The latter part can be made proportional to i_i by a parameter with the dimension of impedance. This is a correlation impedance satisfying the equation

$$e_i = e_n + i_i Z_{cor} \qquad 8.12$$

Treating the noise current(i_i) in a similar way yields a correlation admittance.

$$i_i = i_n + e_i Y_{cor} \qquad 8.13$$

This gives us two noise sources, i_n and e_n, that are uncorrelated. These noise sources and their correlation impedance or admittance with a source termination are shown schematically in Figure 8-4. Since the two noise sources now are uncorrelated, the total noise power spectral density entering the two-port is found to be

$$\left\langle |i_{tot}|^2 \right\rangle = \left\langle |i_s|^2 \right\rangle + \left\langle |i_n|^2 \right\rangle + \left\langle |e_n|^2 \right\rangle |Y_s + Y_{cor}|^2 \qquad 8.14$$

where i_s is the noise current generated by the thermal temperature of the signal source conductance. The actual values of the mean currents and voltages are not known but can be expressed in equivalent resistances or conductances, bandwidths, and temperatures, as given in Equations 8.1 and 8.2,

$$e_n = r_n T_o \qquad \text{and} \qquad i_i = g_n T_o \qquad\qquad 8.15$$

or

$$e_i = R_n T_o \qquad \text{and} \qquad i_n = G_n T_o \qquad\qquad 8.16$$

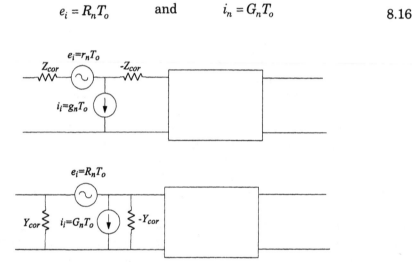

Figure 8-3 *Two models of the input referenced noise sources and correlation impedances and admittances.*

Assuming the signal source and two-port are in thermal equilibrium at the same temperature, the currents and voltages can be expressed by conductances and resistances

$$G_{tot} = G_s + G_n + R_n |Y_s + Y_{cor}|^2 \qquad\qquad 8.17$$

where G_s is the signal source conductance, G_n is the noise-current source Norton conductance, and R_n is the noise-voltage source Thevinen resistance. Here we assume that Y_s and Y_{cor} are constant over the noise bandwidth of interest (sometimes called spot noise bandwidth).

These equations describe a noisy two-port's response to an excitation signal. A simple extension can be used to address the noise in networks with more than two ports. We have included equivalent noise voltage and current sources to account for the noise generated in the two-port network. Thermal noise that does not vary with frequency is called *white noise*. Semiconductors also have a noise power that follows a 1/f function at low frequencies.

Example 8.1: Find the noise voltage, current, and power of a 50 ohm resistor at 290 degrees kelvin in a 1 hertz bandwidth. Using Equation 8.1, the open-circuit, mean-squared voltage of the 50 ohm resistor is

$$\langle |e_n|^2 \rangle = 4kTRB = 4(1.37\text{E}-23)(290)(50)(1)$$

$$= 7.95\text{E}-19$$

$$\langle |i_n|^2 \rangle = 4kTGB = 4(1.37\text{E}-23)(290)(0.02)(1)$$

$$= 3.18\text{E}-22$$

$$P_n = kTB = (1.37\text{E}-23)(290)(1)$$

$$3.97\text{E}-21 = -174 \text{ dB}$$

The noise power coming from a room temperature load is –174 dBm/Hz.

8.3   Noise Power Waves

So far we have represented noise by voltage and current sources. We can also represent noise by a generator of noise power signals [1, 7]. This theory represents noise as power waves such as the a and b waves in S-parameter analysis. A noisy circuit is modeled as a noiseless circuit with noise wave generators instead of noise current and noise voltage sources on the input/output. We can derive the noise power waves from the equivalent voltage and current sources [3].

Suppose there is a noise-voltage source e_i and a noise-current source i_i on the input of a two-port network, as shown in Figure 8-2(c). Assuming these are stationary stochastic processes, two new noise sources are defined as an incident noise wave and a departing noise wave given by

$$a_n = \frac{e_i + Z_v i_i}{2\sqrt{\text{Re}\{Z_v\}}} \qquad\qquad 8.18$$

$$b_n = \frac{e_i - Z_v^* i_i}{2\sqrt{\text{Re}\{Z_v\}}} \qquad\qquad 8.19$$

where Z_v is an arbitrary complex normalization impedance. The strength of a_n and b_n defines temperatures T_a and T_b

$$T_a = \frac{\langle |a_n|^2 \rangle}{k\Delta f}$$

8.20

and

$$T_b = \frac{\langle |b_n|^2 \rangle}{k\Delta f}$$

8.21

respectively where the brackets <> represent the time average of the quantity inside. There is also a temperature associated with the correlation between the a and b wave.

$$T_c = \frac{\langle |a_n^* b_n| \rangle}{k\Delta f}$$

8.22

According to the definition of scattering parameters, the departing waves, b_i ($i = 1, 2, ..., n$) of a linear N-port are related to all the incident waves a_j ($j = 1, 2, ..., n$) by the scattering coefficients S_{ij}. In a noisy N-port, the inherent noise sources are assumed to be independent of the incident waves a_j. An equivalent circuit of a noiseless two-port and two noise wave generators, a_n and b_n, is shown in Figure 8-4. A noisy amplifier is modeled as a noiseless amplifier, and the two noise waves are injected by ideal directional couplers.

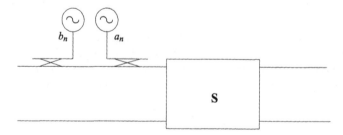

Figure 8-4 *A noiseless two-port with noise waves a_n and b_n representing the internally generated noise.*

The system of linear equations for the two-port is

$$\begin{bmatrix} b_1 - b_n \\ b_2 \end{bmatrix} = \begin{bmatrix} S_{11} & S_{12} \\ S_{21} & S_{22} \end{bmatrix} \begin{bmatrix} a_1 + a_n \\ a_2 \end{bmatrix} \qquad 8.23$$

This set of linear equations can characterize the noise performance of a linear two-port circuit. Two complex power waves are used instead of two complex voltage and/or current sources on the terminals to account for the internal noise generated in a circuit. This equation is more applicable to modern microwave circuit analysis because the nature of the noise sources is similar to the excitation sources used in S-parameter design and analysis methods.

Example 8.2: A noisy two-port is connected to a source termination with a reflection coefficient of G_s. Find the noise waves leaving the output of the two-port. Beginning with Equation 8.23, we set

$$a_2 = 0$$

and a_1 is equal to the noise leaving the source termination. Equation 8.23 is a system of linear equations.

$$b_1 - b_n = (a_1 + a_n)S_{11}$$

$$b_2 = (a_1 + a_n)S_{21}$$

The second equation is solved for $a_1 + a_n$.

$$a_1 + a_n = \frac{b_2}{S_{21}}$$

Then we can solve for b_2.

$$b_2 = b_n \frac{S_{21}}{S_{11}}$$

8.4   Noise Figure

Noise figure measures the amount of noise, thermal or otherwise, added by circuits and systems in the signal path. It is an important design criterion for some circuits. The noise source can be the thermal vibrations of current carriers in a lossy element, the random generation of hole-electron pairs in a semiconductor, or the result from current consisting of discrete charges. Other noise sources can arise from nonlinear effects or electromagnetic interference, which are not included in the characterization of a transistor's noise properties.

Noise figure is defined as a two-port figure of merit pertaining to the amount of noise added by the circuit. To remove the effects of gain or loss in the signal path by the two-port, noise factor is defined as

$$F = \frac{\text{signal-to-noise ratio at the input}}{\text{signal-to-noise ratio at the output}} \qquad 8.24$$

Defining a circuit's noise performance as signal-to-noise ratios removes any signal gain or loss from the actual noise factor value. The result is a measure of how much noise has been added by the circuit. When the ratio is converted to decibels, the measure is usually referred to as *noise figure*.

The noise at the output of an amplifier is the amplified thermal noise of the signal source resistance plus the noise produced by the amplifier. The standard definition of noise figure requires that the source resistance and the load circuit be in thermal equilibrium at a standard temperature of 290° K. It is common to have the signal source represented by another temperature, as with a radiometer or satellite receiver. Noise temperature is a measure of the noise characteristic of a two-port that relates the amount of extra noise added to the signal path in terms of temperature.

$$T = \frac{290(1-F)}{290} \qquad 8.25$$

It is a more representative figure of merit for microwave receiver applications. When the noise sources are both on the input, gain or loss in the two-port will not affect the noise figure. If the two-port is noiseless, the signal-to-noise ratio is the same before and after the two-port. When all the noise sources are on the input, the noise factor is defined as

$$F = \frac{\text{total input noise power to network}}{\text{thermal noise power from the source}} \qquad 8.26$$

Since the noise sources are on the two-port's input, the noise figure is the ratio of total noise input versus noise due to the signal source impedance or conductance. In terms of noise currents

$$F = \frac{\left\langle |i_{tot}|^2 \right\rangle}{\left\langle |i_s|^2 \right\rangle} = \frac{G_{tot}}{G_s} \qquad 8.27$$

From Equations 8.17 and 8.27 the noise figure can be written as a function of source admittance (Y_s), correlation admittance (Y_{cor}), noise current source conductance (G_n) and the noise voltage source resistance (R_n)

$$F = 1 + \frac{1}{G_s}\left(G_n + R_n|Y_s + Y_{cor}|^2\right) \qquad 8.28$$

where G_s is the real part of the signal source admittance. Using Equation 8.31, the noise factor achieves the lowest value at

$$F_{min} = 1 + 2\left\{R_n G_{cor} + \sqrt{R_n G_n + \left(R_n G_{cor}\right)^2}\right\} \qquad 8.29$$

where $Y_{cor} = G_{cor} + jB_{cor}$. The minimum occurs at a signal source admittance of

$$B_{cor} + B_{opt} = 0 \qquad 8.30$$

and

$$G_{opt} = \sqrt{\frac{G_n}{R_n} + G_{cor}^2} \qquad 8.31$$

which yields the admittance of the signal source that causes the lowest noise figure, $Y_{opt} = G_{opt} + jB_{opt}$.

Equations 8.29 through 8.31 are the well-known minimum noise figure and optimum noise source match given in one form or another in most data sheets for low noise transistors. As shown by Equation 8.28, four independent variables are needed to define the noise figure at any source admittance. F_{min} and the real and imaginary part of the optimum noise figure source admittance are three of the characteristic values. Another quantity commonly given in the data sheets is R_n, the noise-voltage Thevinen resistance. It is convenient to use these four quantities in one equation for noise factor [2].

$$F = F_{min} + \frac{R_n}{G_s}|Y_s + Y_{cor}|^2 \qquad 8.32$$

Equation 8.32 is the expression for noise factor as a function of these four characteristic values and the signal source admittance. A dual expression using the noise voltages and impedances would yield the equation for a noise factor of

$$F = F_{\min} + \frac{g_n}{R_s}\left|Z_s - Z_{opt}\right|^2 \qquad\qquad 8.33$$

if the fourth characteristic parameter is g_n. The noise parameters g_n, r_n, and Z_{cor} in Figure 8-3(b) are related to G_n, R_n, and Y_{cor} in Figure 8-3(a) by the following equations

$$g_n = G_n + R_n \left|Y_{cor}\right|^2 \qquad\qquad 8.34$$

$$R_n = r_n + g_n \left|Z_{cor}\right|^2 \qquad\qquad 8.35$$

$$r_n = \frac{G_n}{\left|Y_{cor}\right|^2 + \left(\dfrac{G_n}{R_n}\right)} \qquad\qquad 8.36$$

$$G_n = \frac{r_n}{\left|Z_{cor}\right|^2 + \left(\dfrac{r_n}{g_n}\right)} \qquad\qquad 8.37$$

$$Z_{cor} = \frac{Y_{cor}^*}{\left|Y_{cor}\right|^2 + \left(\dfrac{G_n}{R_n}\right)} \qquad\qquad 8.38$$

$$Y_{cor} = \frac{Z_{cor}^*}{\left|Z_{cor}\right|^2 + \left(\dfrac{r_n}{g_n}\right)} \qquad\qquad 8.39$$

Using the definition of reflection coefficient, we can write Equation 8.32 in terms of source reflection coefficient Γ_s and optimum source reflection coefficient Γ_{opt} instead of admittance [8]

$$F = F_{\min} + \frac{R_n}{G_s}\left|\frac{1-\Gamma_s}{1+\Gamma_s} - \frac{1-\Gamma_{opt}}{1+\Gamma_{opt}}\right|^2 \qquad\qquad 8.40$$

$$F = F_{\min} + \frac{R_n}{G_s}\left|\frac{(1-\Gamma_s)(1+\Gamma_{opt}) - (1-\Gamma_{opt})(1+\Gamma_s)}{(1+\Gamma_s)(1+\Gamma_{opt})}\right|^2 \qquad\qquad 8.41$$

$$F = F_{\min} + \frac{4\,R_n}{Z_o}\,\frac{\left|\Gamma_s - \Gamma_{opt}\right|^2}{\left(1 - \left|\Gamma_s\right|^2\right)\left|1 + \Gamma_{opt}\right|^2} \qquad 8.42$$

where Γ_s is the source reflection coefficient, and Γ_{opt} is the source reflection coefficient yielding the minimum noise figure.

Figure 8-4 shows a two-port with two input referenced noise waves, an incoming wave a_n and an outgoing wave b_n. It has been shown [1, 4] that an impedance Z_v can be found that leaves a_n and b_n uncorrelated, satisfying the condition

$$\left\langle a_n^* b_n \right\rangle = 0 \qquad 8.43$$

According to Figure 8-4, a source impedance connected to the input would have a reflection coefficient of

$$\Gamma_s = \frac{Z_s - Z_v}{Z_s + Z_v^*} \qquad 8.44$$

Since all noise sources are on the input, the resultant noiseless two-port does not affect the noise figure. The two-port can be replaced by a matched load to calculate the noise figure. The noise power incident on the two-port due to the internal noise sources is

$$P_{inc} = \left\langle \left| a_n + \Gamma_s b_n \right|^2 \right\rangle \qquad 8.45$$

Since a_n and b_n are uncorrelated, Equation 8.45 can be written as

$$P_{inc} = \left\langle \left| a_n \right|^2 \right\rangle + \left| \Gamma_s \right|^2 \left\langle \left| b_n \right|^2 \right\rangle \qquad 8.46$$

The noise power from the source impedance at temperature T_o is

$$P_s = k T_o \Delta f \left(1 - \left| \Gamma_s \right|^2 \right) \qquad 8.47$$

The ratio of Equations 8.46 and 8.47 is the excess noise figure.

$$F - 1 = \frac{P_{inc}}{P_s} = \frac{\left\langle \left| a_n \right|^2 \right\rangle + \left| \Gamma_s \right|^2 \left\langle \left| b_n \right|^2 \right\rangle}{k T_o \Delta f (1 - \left| \Gamma_s \right|^2)} \qquad 8.48$$

The noise waves in Figure 8-3 are modeled as thermal noise. When the temperatures are used instead of the mean magnitudes of the noise waves, Equation 8.48 becomes

$$F = 1 + \frac{T_a + T_b|\Gamma_s|^2}{T_o\left(1 - |\Gamma_s|^2\right)}$$ 8.49

which can be written in terms of source and normalization impedance as

$$F = 1 + \frac{T_a}{T_o} + \frac{T_a + T_b}{T_o} \frac{|Z_s - Z_v|^2}{4 \operatorname{Re}\{Z_s\}\operatorname{Re}\{Z_v\}}$$ 8.50

Equation 8.50 is very similar to the equations for noise factor in Equations 8.32 and 8.33. Since T_a and T_b both are positive real numbers, the noise figure reaches the minimum value when Z_s equals Z_v. This shows that the normalization impedance Z_v that satisfies Equation 8.43 is, in fact, the optimum noise source impedance. Comparing Equation 8.50 to Equation 8.29 and bearing in mind that Z_v equals Z_{opt}, the minimum noise factor is

$$F_{\min} = 1 + \frac{T_a}{T_o}$$ 8.51

and r_n is given by

$$r_n = \frac{R_n}{50} = \frac{T_a + T_b}{T_o 4 \operatorname{Re}\left\{\dfrac{1}{Z_{opt}}\right\}}$$ 8.52

These equations represent the noise performance of a circuit when the two noise waves, one incoming and one outgoing, are uncorrelated.

The requirement to normalize the S-parameters by the optimum noise match poses practical problems. First, the calculations to convert to the new S-parameters with a complex normalization impedance are lengthy. More importantly, the technique requires prior knowledge of the optimum noise source impedance. This presents a problem when an imbedding circuit changes the optimum noise source impedance. When the normalization impedance is not the optimum noise match, the two noise waves (a_n and b_n) are partially correlated. Meys [6] introduced the complex temperature coefficient $T_c e^{j\phi}$ given by the relation

$$\left\langle a_n^* b_n \right\rangle = k\Delta f T_c e^{j\phi} \qquad 8.53$$

Using this new temperature coefficient, any set of noise and S-parameters normalized to an arbitrary impedance can be used. In this discussion, it will be assumed that the normalization impedance is 50 ohms. Most transistor data sheets have tabulated these parameters, but the technique is applicable to any normalization impedance. If a_n and b_n are correlated, Equation 8.45 becomes

$$P_{inc} = \left\langle |a_n|^2 \right\rangle + 2\mathrm{Re}\left\{ \Gamma_s \left\langle a_n^* b_n \right\rangle \right\} + |\Gamma_s|^2 \left\langle |b_n|^2 \right\rangle \qquad 8.54$$

The noise figure is found to be

$$F = 1 + \frac{T_a + 2\,\mathrm{Re}\left\{ \Gamma_s\, T_c e^{j\phi_c} \right\} + T_b |\Gamma_s|^2}{T_o\left(1 - |\Gamma_s|^2\right)} \qquad 8.55$$

By comparing Equation 8.42 to Equation 8.55, we can find the three noise temperatures from the four commonly used noise parameters $F_{\min}$, R_n and optimum match Γ_{opt} [12].

$$T_a = \left(F_{\min} - 1\right)T_o + \frac{4\,r_n T_o |\Gamma_{opt}|^2}{|1 + \Gamma_{opt}|^2} \qquad 8.56$$

$$T_b = \frac{4\,r_n T_o}{|1 + \Gamma_{opt}|^2} - \left(F_{\min} - 1\right)T_o \qquad 8.57$$

$$T_c = \frac{4\,r_n\,T_o\,|\Gamma_{opt}|}{|1 + \Gamma_{opt}|^2} \qquad 8.58$$

$$\phi_c = \pi - \arg\left(\Gamma_{opt}\right) \qquad 8.59$$

Equations 8.56 through 8.59 are used to calculate the noise temperatures from the more common noise figure parameters given for microwave transistors. When noise temperatures are known, the minimum noise figure, noise resistance, and optimum noise match can be found by solving Equations 8.56 through 8.59.

$$\arg\left(\Gamma_{opt}\right) = \pi - \phi_c \qquad 8.60$$

$$|\Gamma_{opt}| = \frac{T_a + T_b}{2|T_c|} - \sqrt{\left(\frac{T_a + T_b}{2|T_c|}\right)2 - 1}$$ 8.61

$$r_n = \frac{|T_c| \; \left|1 + \Gamma_{opt}\right|^2}{|\Gamma_{opt}| 4T_o}$$ 8.62

$$F_{min} = 1 + \frac{T_a - T_b}{2T_o} - \frac{|T_c|\left(|\Gamma_{opt}|^2 - 1\right)}{2 \, T_o \; |\Gamma_{opt}|}$$ 8.63

When using the concept of a complex temperature, the general equations for noise figure are available with any two-port normalized to an impedance where the two noise waves a_n and b_n are correlated. This form is more applicable than using the uncorrelated noise temperatures because most S-parameters are normalized to 50 ohms.

Noise waves are similar to noise temperatures. In microwave applications, as in radiometry and communications, where noise temperature is common, the noise wave technique can be used to find noise temperatures directly from an amplifier's four characteristic noise variables. With a slight modification of Equation 8.55, the noise temperature of the amplifier can be written as

$$T = \frac{T_a + 2 \; \text{Re}\left\{\Gamma_s \, T_c e^{j\phi_c}\right\} + T_b|\Gamma_s|^2}{\left(1 - |\Gamma_s|^2\right)}$$ 8.64

Example 8.3: Find the noise temperature and noise factor of a two-port with a F_{min} of .52 dB, $\Gamma_{opt} = 0.64 \; \angle -32.6°$, and $r_n = 1.2$ when the amplifier is connected to a $52 + j12$ ohm source.

$$\Gamma_s = \frac{52 + j12 - 50}{52 + j12 + 50} = .033 + j.114$$

$$F_{min} = 10^{0.052} = 1.127$$

$$F = 1.127 + 4(1.2)\frac{|\Gamma_s - 0.64@-32.6|}{\left(1 - |\Gamma_s|^2\right)|1 + 0.64@-32.6|^2} = 2.04$$

$$T_a = (1.127 - 1)290 + \frac{4 \; (1.2)(290)|0.64|^2}{|1 + 0.64@-32.6|^2} = 266$$

$$T_b = \frac{4\,(1.2)\,290}{|1+0.64@{-}32.6|2} - (1.127-1)(290) = 523$$

$$T_c = \frac{4\,(1.2)\,290\,|0.64|}{|1 + 0.64@{-}32.6|^2} = 359$$

$$\phi_c = \pi - \arg(0.64@{-}32.6) = 3.71 \text{ rad}$$

$$T = \frac{T_a + 2\,\mathrm{Re}\{(0.64@{-}32.6)T_c e^{j\phi_c}\} + T_b|0.64|^2}{(1-|0.64|^2)} = 301$$

$$F = 1 + \frac{301}{290} = 2.04$$

8.5   Measuring Noise Figure

The measurement of noise figure is an integral part of the design process. Manufacturers usually provide measurement information on these noise parameters but the noise figure of the entire amplifier circuit may have to be determined. The noise figure can be measured using two or more noise sources, which can be resistors at different temperatures, ionized gas, or a noisy diode. The noise sources are connected to the input of the amplifier, and the amplifier's output noise power is measured by a sensitive power meter.

Figure 8-5(a) shows a simple circuit with a noise source at temperature T_s connected to an amplifier. The amplifier has a gain of G_a and adds some noise to the system N_a. The total noise output from the circuit is

$$N_o = N_a + kT_s BG_a \qquad\qquad 8.65$$

Equation 8.65 has one unknown, i.e., the noise added by the amplifier N_a. The equivalent noise temperature of the amplifier is

$$T_a = \frac{N_a}{kBG_a} \qquad\qquad 8.66$$

yielding a noise figure of

$$F = \frac{N_a}{290\,kBG_a} - 1 \qquad\qquad 8.67$$

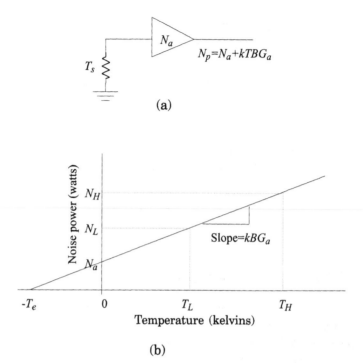

Figures 8-5 *(a) A simple noise measurement block diagram and (b) a plot of the noise power measured as a function of the noise temperature of the source.*

The gain and noise figure of an amplifier can be measured with two noise sources at different temperatures. These two quantities are best measured by the same method to minimize potential errors caused by using two independent measurement systems. Two noise sources at temperatures T_L and T_H are connected to the input of the amplifier. The output noise power N_L and N_H are measured with each one of the noise sources. Figure 8-5(b) plots the output noise power as a function of the source noise temperature. The temperature T_H is higher than T_L, which results in a higher measured output noise power (N_H). There are two equations and two unknowns.

$$N_L = N_a + kT_L BG_a \qquad\qquad 8.68$$

$$N_H = N_a + kT_H BG_a \qquad\qquad 8.69$$

Now we can see that Equations 8.65, 8.68, and 8.69 are equations for a line where N_a is the intercept point and kBG_a is the slope.

$$G_a = \frac{N_H - N_L}{kB(T_H - T_L)} \qquad 8.70$$

There will always be a positive intercept point because all circuits will add noise. Figure 8-5(b) illustrates that if a noise source at absolute zero could be obtained, the output noise power would be solely due to the amplifier at point N_a.

Two noise sources are needed to simultaneously measure noise figure and gain. It is common to express the two noise sources in terms of their temperature ratio, or excess noise ratio (ENR). We will also want to express the output noise power as a ratio of the noise power measurements at the two source temperatures. This ratio is called the Y-factor

$$Y = \frac{N_H}{N_L} = \frac{kT_H B}{kT_L B} \qquad 8.71$$

where T_H is the high temperature and T_L is the low temperature. The ENR is defined as

$$\text{ENR} = 10 \ \log\left(\frac{T_H - T_L}{290}\right) \qquad 8.72$$

and is expressed in dB. We take the ratio of the two output noise measurements from Equations 8.68 and 8.69

$$Y = \frac{N_H}{N_L} = \frac{N_a + kT_H BG_a}{N_a + kT_L BG_a} \qquad 8.73$$

and solve for the noise figure of the amplifier.

$$Y(N_a + kT_L BG_a) = N_a + kT_H BG_a \qquad 8.74$$

$$N_a(Y - 1) = kBG_a(T_H - YT_L) \qquad 8.75$$

$$\frac{N_a}{kBG_a} = \frac{T_H + YT_L}{Y - 1} \qquad 8.76$$

When $T_L = 290$ kelvins, the noise figure of an amplifier is

$$F = \text{ENR} - 10\log(Y - 1) \ \text{dB} \qquad 8.77$$

A noise source with a calibrated ENR and a low source temperature of 290 K is common in automatic noise measurement equipment. The noise source is usually a diode, which has both an on and an off state. When the diode is off, the noise source is at room temperature. When the diode is on, it creates noise which gives it a high noise temperature.

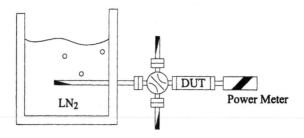

Figure 8-6 *A waveguide noise figure test setup with a cold load and a room temperature load.*

At some frequencies, good solid-state noise sources are not available. For example, a termination in liquid nitrogen is commonly used as the cold load. Figure 8-6 shows a vessel partially filled with liquid nitrogen, containing a waveguide and a load. The temperature of the waveguide termination is at the boiling point of liquid nitrogen. Losses in the waveguide leading out of the container will raise the noise temperature by some amount. The power in a wave traveling in the positive x direction in a transmission line with an attenuation function of $\alpha(x)$ and a temperature of $T(x)$ is a linear first-order differential equation.

$$\frac{dP}{dx} = -2\alpha(x)P + 2\alpha(x)kT(x)B \qquad 8.78$$

The general solution to this differential equation is

$$T' = \frac{T + \int_0^l 2\alpha(x)T(x)e^{\int 2\alpha(x)dx}dx}{e^{\int 2\alpha(x)dx}} \qquad 8.79$$

When the transmission line is at a constant temperature of T_T and a total loss of α, the noise temperature at the output of the transmission line is

$$T = \alpha T_L + T_T(1 - \alpha) \qquad 8.80$$

where T_L is the temperature of the termination. When the transmission line has a constant loss-per-unit length of α, but a linear temperature distribution, the noise temperature along the transmission line is

$$T = T_L + (T_T + T_L)\left(1 - e^{\left(\frac{\alpha-1}{\alpha^2}\right)}\right) \qquad 8.81$$

The differential equation can be solved for other transmission line loss. The temperature distribution has been solved by Stelzried [11].

Care must be exercised when designing a noise figure test such as the one shown in Figure 8-5(a). We have seen that the source reflection coefficient affects the noise power delivered to the device under test (DUT). Any experiment using a noise source as the one described above assumes that the ENR is independent of the DUT input match. Equation 8.45 shows that the highest accuracy is obtained when the noise source has a reflection coefficient of zero. We have to be careful to ensure that the noise source has a reflection coefficient as close to zero as possible. Lani [5] has found an efficient method of measuring the optimum noise match Γ_{opt}, minimum noise figure F_{min}, and noise resistance R_n based on measurements of noise figure at four or more different source reflection coefficients (Γ_s).

Example 8.4: Two hot loads with temperatures of 1200 K and 4500 K are used in a noise figure test set. There is 1.5 dB of loss in a transmission line at 290 K between the noise source and the device under test. Combine these two noise sources with a room temperature load, 290 K, and find all the possible ENRs that this test setup can produce. First, we calculate the temperatures of the noise sources including the loss of the transmission line using Equation 8.80.

$$\alpha = 10^{-.15} = 0.7079$$

$$T_2 = (0.7079)1200 = (1 - 0.7079)290 = 934K$$

$$T_3 = (0.7079)(4500) + (1 - 0.7079)290 = 3186K$$

The noise temperature of the room temperature load is not affected by the loss between the load and the device under test. Three possible ENR values can be obtained with these three loads.

$$ENR = 10\ \log_{10}\left(\frac{765 - 290}{290}\right) = 2.14\ \text{dB}$$

$$\text{ENR} = 10 \ \log_{10}\left(\frac{3101 - 290}{290}\right) = 9.86 \text{ dB}$$

$$\text{ENR} = 10 \ \log_{10}\left(\frac{3101 - 765}{290}\right) = 9.06 \text{ dB}$$

Example 8.5: An amplifier is tested with a noise source having an ENR of 12.3 dB. The noise power of the output is measured over a 10 MHz bandwidth and is –110 dBm with the noise source on and –118 dBm with the noise source off. What is the Y-factor, gain and noise figure of the amplifier?

$$-110 \text{ dBm} = 1.00\text{E} - 11$$

$$-118 \text{ dBm} = 1.58\text{E} - 12$$

$$Y = \frac{1.00\text{E} - 11}{1.58\text{E} - 12} = 6.396$$

$$G_a = \frac{N_H - N_L}{kB\left(10^{ENR/10}\right)290} = \frac{(1.00\text{E} - 11) - (1.58\text{E} - 12)}{(1.374\text{E} - 23)(10E6)\left(10^{1.23}\right)290} = 12.44$$

$$F = 10 \ \log_{10}(6.396 - 1) = 7.25 \text{ dB}$$

8.6   Summary

This chapter described the origin of noise and noise figure. All practical circuits add noise to the signal passing through them. Passive circuits add noise due to resistive losses in the signal path. These losses will add noise if the circuit is at a finite temperature. Active circuits will add noise due to random current and voltage effects within the active circuit. Many systems are sensitive to noise added by the circuits designed to amplify, filter, or process signals flowing through them. It is important to be able to characterize the noise properties of an amplifier. Furthermore, it may be necessary to design an amplifier to add as little noise as possible.

Section 8.2 introduced the basic noise model as small random voltage and/or current fluctuations. The noise generated in a one-port requires only one equivalent source of noise. A two-port requires at least two sources of noise to fully model the noise generated within the network. The two noise sources are correlated, some of the noise in these two sources is common between them and some is not. This yields three different quantities: two independent noise factors, one from each source, and one that is correlated. We can use an impedance or an admittance to describe the amount of correlation between the two noise sources.

Section 8.3 explained noise power waves as an alternate method to describe the equivalent noise sources described in Section 8.2. Noise power waves are similar to the power waves that are used to define S-parameters. They are directly applicable to the linear equations describing S-parameter networks.

Section 8.4 defines the noise figure and noise temperature of a circuit. Instead of describing noise as a collection of noise sources, we derived expressions for noise figure as a function of minimum noise figure, optimum noise impedance, and a noise scaling factor. These three parameters are commonly published in transistor data sheets and serve as a starting point for designing low noise amplifiers.

Finally, Section 8.5 showed how to measure noise figure by using one or more noise sources. These noise sources are modeled as a source at a certain temperature. The gain and noise figure can be measured by using two noise sources each at a different temperature. By switching between the noise sources, the output noise power is measured in each state. An automatic noise figure measurement system uses a noise source with a calibrated excess noise ratio (ENR).

8.7 Problems

8.1 Find the noise voltage, current and power emitted by a 10, 50, 350, and 10 kohm resistor at room temperature.

8.2 Given i_{n1}, i_{n2} and the Y-parameters of a two-port network, find the equations for e_i and i_i.

8.3 Given e_i, i_i, and the two-port Z-parameters, find the equations for e_{n1} and e_{n2}.

8.4 What is the noise power for a resistor at liquid nitrogen, 70 K, and the clear sky temperature of 4 K?

8.5 Find an expression of the noise waves leaving Port 1 as a function of a_n and b_n when Port 2 is terminated with a load having a reflection coefficient of Γ_L.

8.6 Find the noise figure of a two-port device that has a source termination of 50 Ω = Z_o when the minimum noise figure is 2.3 dB, R = 2.3 and Γ_{opt} = 0.3 + j0.6.

8.7 Find the noise temperatures T_a, T_b, and T_c using values given in Problem 8.6, then find the noise temperature of the two-port.

8.8 A noise source can be switched on and off. It has a noise temperature of 2400 K when it is on, and it is at room temperature when it is off. Find the ENR of the noise source.

8.9 A microwave receiver has 46 dB of gain, a bandwidth of 4.8 MHz and a noise figure of 2.5 dB. What is the output noise power of the receiver when it is connected to an antenna with a noise temperature of 350 K?

8.10 The receiver in Problem 8.9 is tested with a noise source with an ENR of 13.4 dB and the low temperature of the noise source is 290 K. What noise power will be measured with the noise source on and off?

8.8 References

1. Bauer, H. and H. Rothe. "Die Äquivalenten Rauschvierpole als Wellenvierpole." *Archive der Elektrischen Übertragung* 10, no. 6 (1956): 241-252.

2. Haus, H. A. "Representation of Noise in Linear Two-Ports." *Proceedings of the IRE* 48, no. 1 (1960): 69-74.

3. Hecken, R. P. "Analysis of Linear Noisy Two-Ports Using Scattering Waves." *IEEE Transactions on Microwave Theory and Techniques* MTT-29, no. 10 (1981): 997-1003.

4. Hirano, K. and S. Kenama. "Matrix Representation of Noise Figures and Noise Figure Charts in Terms of Power Wave Variables." *IEEE Transactions on Microwave Theory and Techniques* vol. MTT-16, no. 9 (September 1968): 692-698.

5. Lani, R. Q. "The Determination of Device Noise Parameters." *Proceedings of the IEEE* 57, no. 8 (1969): 1461-1462.

6.   Meys, R. P. "A Wave Approach to the Noise Properties of Linear Microwave Devices." *IEEE Transactions on Microwave Theory and Techniques* vol. MTT-26, no. 1 (1978): 34-37.

7.   Penfield, P. "Wave Representation of Amplifier Noise." *IRE Transactions on Circuit Theory* CT-9, (1962): 84-86.

8.   Poole, C. R. and D. K. Paul. "Optimum Noise Measure Terminations for Microwave Transistor Amplifiers." *IEEE Transactions on Microwave Theory and Techniques* vol. MTT-33, no. 11 (1985): 1254-1257.

9.   Rothe, H. and W. Dahlke. "Theory of Noisy Fourpoles." *Proceedings of the IRE* 44, no. 6 (1956): 811-818.

10. Siegman, A. E. "Thermal Noise in Microwave Systems." Parts I and II. *Microwave Journal*, no. 3 (1961): 81-90; no. 4 (1961): 66-73.

11. Stelzried, C. T. "Temperature Calibration of Microwave Thermal Sources." *IEEE Transactions on Microwave Theory and Techniques* vol. MTT-13, no. 1 (1965): 128-129.

12. Withington, S. "Scattered Noise Waves in Microwave and mm-Wave Networks." *Microwave Journal*, no. 6 (1989): 169-178.

Low Noise Amplifier Design

9.1   Introduction

In Chapter 8, we introduced noise and noise figure in mathematical terms. In this chapter, we will show how this information applies to transistor amplifiers. Two-port S-parameters and noise characteristics are published by microwave manufacturers to ensure that the transistor can be easily utilized in an amplifier design. We have introduced a mathematical representation of noise created by a two-port. The noise added by the two-port is characterized by a few quantities that are essential to the two-port and the input reflection coefficient. Some transistor data sheets show noise parameters for a particular transistor at several frequencies.

Many applications, such as communication and radar receivers, require amplifiers that add a minimal amount of noise to the receiver signal. It is necessary to design amplifiers with a specified noise figure to gurantee the proper operation of the receiver. If the amplifier is narrowband, we may be able to design the input matching circuit in a way that the source reflection coefficient corresponds to the minimum noise figure source impedance. We know this will yield an amplifier with the lowest possible noise figure for a particular transistor. However, the optimum noise figure reflection coefficient may lie in the region of potential instability. Moreover, we may want to trade a little noise for a higher gain. This chapter shows how to design low noise amplifiers by finding the constant noise figure circles, using more than one transistor amplifier stage and setting the proper bias point of the transistor.

Section 9.2 discusses constant noise figure circles that are similar to the constant gain circles derived in Chapter 6. The noise figure of an amplifier depends on the noise characteristics of the transistor and the input reflection coefficient. Transistor data sheets include the optimum noise figure, optimum noise figure input match, and the noise correlation resistance at several frequencies. Circles of constant noise figure can be plotted on the input reflection coefficient plane. We can also plot the input stability circle and constant gain circles on the same Smith chart. A noise figure can be chosen to yield a particular noise figure and gain.

Section 9.3 shows how to calculate the noise performance of two or more circuits connected in cascade. Each circuit amplifies (or attenuates) the signal and noise at its input and adds its noise to the signal path. As the gain accumulates through the amplifier chain, the added noise affects the signal-to-noise less and less. We will show how to calculate the noise figure as each amplifier is added to the cascade. Noise measure is introduced as a means of showing the relative noise performance of an amplifier when used in a system composed of other amplifiers and microwave components.

Section 9.4 shows the relation between the transistor bias point and the noise performance of an amplifier. We will also discuss the dynamic range as related to low noise amplifiers and amplifier distortion as well as the dynamic range of an amplifier, which is bound by thermal noise on the low end and the distortion of a high power signal on the upper end.

9.2 Constant Noise Figure Circles

In the previous chapter, we stated that a noisy two-port has a minimum noise figure, an optimum noise match, and a noise resistance. This section will show how this applies to transistors and amplifiers. Characteristic noise parameters can be found in the data tables for the low noise transistor. In Chapter 8, we derived the equation of noise figure as a function of the noise characteristics.

$$F = F_{min} + \frac{4\,R_n}{Z_o}\,\frac{|\Gamma_s - \Gamma_{opt}|}{(1 - |\Gamma_s|^2)|1 + \Gamma_{opt}|^2} \qquad 9.1$$

When $\Gamma_s = \Gamma_{opt}$, the right side of Equation 9.1 is equal to zero, and the noise figure is equal to F_{min}. R_n is always greater than zero. When Γ_s is not equal to Γ_{opt}, the right side of Equation 9.1 is always greater than zero. The noise figure will always be greater than F_{min}. We can view the correlation resistance R_n as a scaling factor that controls the effect of Γ_s on noise figure.

We can rearrange Equation 9.1 to collect only the terms that depend on the source reflection coefficient to the right side of the equation.

$$(F - F_{min})\frac{\left|1 + \Gamma_{opt}\right|^2 Z_o}{4R_n} = \frac{\left|\Gamma_s - \Gamma_{opt}\right|}{(1 - \left|\Gamma_s\right|^2)} \qquad 9.2$$

The left side of Equation 9.2, which we will call N, does not depend on the source reflection coefficient.

$$N = \frac{Z_o(F - F_{min})\left|1 + \Gamma_{opt}\right|^2}{4R_n} \qquad 9.3$$

It is possible to map contours of constant noise figure greater than F_{min} on the source reflection coefficient plane. Equation 9.2 can be rewritten as

$$\left(\Gamma_s - \Gamma_{opt}\right)\left(\Gamma_s - \Gamma_{opt}\right)^* = N - N\left|\Gamma_s\right|^2 \qquad 9.4$$

$$\left|\Gamma_s\right|^2(1 + N) + \left|\Gamma_{opt}\right|^2 - 2\operatorname{Re}\{\Gamma_s\Gamma_{opt}\} = N \qquad 9.5$$

Multiplying both sides of Equation 9.5 by $N + 1$ yields

$$\left|\Gamma_s\right|^2(N+1)^2 + \left|\Gamma_{opt}\right|^2(N+1) - 2(N+1)\operatorname{Re}\{\Gamma_s\Gamma_{opt}\} = N(N+1) \qquad 9.6$$

$$\left|\Gamma_s(N+1) - \Gamma_{opt}\right|^2 = N^2 + N\left(1 - \left|\Gamma_{opt}\right|^2\right) \qquad 9.7$$

or

$$\left|\Gamma_s - \frac{\Gamma_{opt}}{1+N}\right|^2 = \frac{N^2 + \left(1 - \left|\Gamma_{opt}\right|^2\right)}{1 + N^2}. \qquad 9.8$$

Equation 9.8 describes circles of constant noise figure on the complex Γ_s plane centered at

$$C_f = \frac{\Gamma_{opt}}{N+1} \qquad 9.9$$

with a radius of

$$R_f = \frac{\sqrt{N\left(N+\left(1-\left|\Gamma_{opt}\right|^2\right)\right)}}{N+1} \tag{9.10}$$

Since N and F are closely related, contours of constant noise figure are circles in the input source reflection coefficient plane.

Table 9-1 *Noise figure parameters of the Kukje KH1031-C02 at a drain current of 15 mA and voltage of 2 V.*

Frequency	F_{min}	$\left|\Gamma_{opt}\right|$	Arg(Γ_{opt})	R_n	G_a
4 GHz	0.35 dB	0.60	51.0°	12 Ω	16 dB
8 GHz	0.45 dB	0.46	101.5°	7 Ω	12 dB
12 GHz	0.70 dB	0.43	165.2°	2 Ω	10 dB

Most manufacturers and distributors of small signal transistors publish the noise parameters in their data sheets. Table 9-1 shows the noise parameters for the low noise transistor Kukje KH1031-C02 at several frequencies. When designing low noise amplifiers, these noise parameters constitute another data set that is used along with the two-port S-parameters. Each transistor will have a unique set of noise figure parameters measured by the manufacturer. Notice that the minimum noise figure increases as the frequency increases, a behavior which is characteristic for all transistors. Finally, the optimum noise figure match is very reflective at low frequencies and approaches 50 ohms as the frequency increases. The constant noise temperature circles have the same form as Equation 9.2 with

$$T - T_{min}\left(\frac{|1+\Gamma_{opt}|^2 Z_o}{290 \, 4 \, R_n}\right) = \frac{\left|\Gamma_s - \Gamma_{opt}\right|}{290\left(1-\left|\Gamma_s\right|^2\right)} \tag{9.11}$$

which results in a similar family of circles and replacing N with

$$N = \frac{Z_o(T - T_{min})|1+\Gamma_{opt}|2}{1160 \, R_n} \tag{9.12}$$

Example 9.1: Plot the constant 0.6, 0.8, 1.0, and 1.5 dB noise figure circles on the source impedance plane for the Kukje KH1031-C02 FET at 8 GHz.

$$F_{min} = 1.1092$$

$$F_1 = 1.1482$$

$$N_1 = \frac{Z_o(1.1482 - 1.1092)\left|1 + 0.46@101.5^\circ\right|^2}{4(7.0)} = 0.0716$$

$$C_{f1} = \frac{0.46@101.5^\circ}{0.0716 + 1} = -0.0856 + j0.4207$$

$$R_{f1} = \frac{\sqrt{0.0716\left(0.0716 + \left(1 - |0.46|^2\right)\right)}}{0.0716 + 1} = 0.2828$$

$$F_2 = 1.2023, \quad N_2 = 0.1709$$

$$C_{F2} = -0.07083 + j0.3850, \quad R_{F2} = 0.4151$$

$$F_3 = 1.2589, \quad N_3 = 0.2749$$

$$C_{F3} = -0.0719 + j0.3536, \quad R_{F3} = 0.5014$$

$$F_4 = 1.4128, \quad N_4 = 0.5570$$

$$C_{F4} = -0.0589 + j0.2895, \quad R_{F4} = 0.6375$$

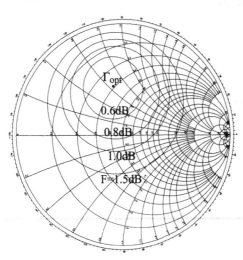

Figure 9-1 *Constant noise figure circles plotted on the source reflection coefficient plane.*

We have seen that the noise figure and gain of an amplifier depend on the input reflection coefficient of the transistor. The optimum noise figure and the conjugate gain matches are not the same, and neither is the family of constant gain and noise figure circles. The combination of gain and noise figure is unlimited. The easiest way to design an amplifier with specific gain and noise figure specifications is to plot the constant gain and noise figure circles on the source reflection coefficient plane. We use the constant available power gain circles given in Equations 6.68 and 6.69 along with the constant noise figure circles given above.

Example 9.2: The S-parameters and noise figure parameters of the Mitsubishi Electronics MGF-4300A transistor at 4 GHz, 2 volts and 7.5 mA are given below. Show the constant gain and noise figure circles for this transistor at this bias point.

$S_{11} = 0.863@-79.1°$ $S_{12} = 0.072@36.5°$ $S_{21} = 3.434@106.2°$
$S_{22} = 0.627@-58.3°$

$F_{min} = 0.46$ dB $R_n = 13.5$ $\Gamma_{opt} = 0.740@62.0°$

Mitsubishi has calculated the stability factor for this FET to be K = 0.338. Since the transistor is only conditionally stable, we must calculate the input and output stability circles.

$\Delta = (0.1632 - j.8474)(0.3295 - j0.5335) - (0.0579 + j0.0428)(-0.9581 + j3.2976)$
$= -0.2016 - j0.5161$

According to the given S-parameters, the stability factor is

$$K = .3420$$

The load stability circle is centered at

$$C_L = -0.8705 + j3.232$$

with a radius of

$$R_L = 2.8706$$

The source stability circle is centered at

$$C_s = -0.1044 + j1.3017$$

and has a radius of

$$R_s = 0.5648$$

We can now calculate input constant noise figure circles.

$$F_{min} = 1.11$$

$$F_1 = 1.2$$

$$N = \frac{50(1.2 - 1.11)\left|1 + 0.740@62.0°\right|^2}{4(13.5)} = 0.183$$

$$C_{f1} = \frac{0.740@62.0°}{0.183 + 1} = 0.29 + j0.55$$

$$R_{f1} = \frac{\sqrt{0.183\left(0.183 + \left(1 - |0.74|^2\right)\right)}}{0.183 + 1} = 0.476$$

$$F_2 = 1.3, N_2 = 0.391$$

$$C_{f2} = 0.250 + j0.470, R_{f2} = 0.558$$

$$F_3 = 1.5, N_3 = 0.806$$

$$C_{f3} = 0.192 = j0.362, R_{f3} = 0.6437$$

$$F_4 = 2.0, N_4 = 1.8443$$

$$C_{f4} = 0.1221 + j0.2297, R_{f4} = 0.8794$$

We then calculate circles of constant available power gain. The maximum available gain is not defined since the device is potentially unstable. The maximum stable gain is

$$MSG = \left|\frac{S_{12}}{S_{21}}\right| = 47.69 = 16.8 \text{ dB}$$

The constant source 16.8 dB available gain circle is located at

$$C_{a1} = \frac{G_d\left(S_{11}^* - \Delta^* S_{22}\right)}{|S_{21}|^2 + G_d\left(|S_{11}|^2 - |\Delta|^2\right)} = -0.0667 + j0.8318$$

and has a radius of

$$R_{a1} = \frac{|S_{21}|\sqrt{|S_{21}|^2 - 2G_d K|S_{21}S_{12}| + G_d^2|S_{12}|^2}}{\left||S_{21}|^2 + G_d\left(|S_{11}|^2 - |\Delta|^2\right)\right|} = 0.4141$$

The constant 15, 12, 10 and 8 dB source gain circles are

$$C_{a2} = -0.0564 + j0.7029, \quad R_{a2} = 0.4568$$

$$C_{a3} = -0.0387 + j0.4822, \quad R_{a3} = 0.5917$$

$$C_{a4} = -0.0283 + j0.3524, \quad R_{a4} = 0.6921$$

$$C_{a5} = -0.0198 + j0.2470, \quad R_{a5} = 0.7801$$

respectively. Figure 9-2 shows the constant gain and noise figure circles plotted on the input reflection coefficient plane. These circles trace out two sets of contours, helping us trade off between gain and noise figure. An input match that yields the lowest noise figure would have an approximate gain of 16 dB. An amplifier with a gain of 16.8 dB can nevertheless have a very low noise figure if the source match is chosen on the 16.8 dB gain circle that passes nearest to the Γ_{opt} point.

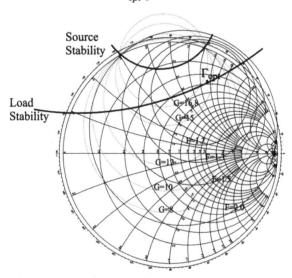

Figure 9-2 *Circles of constant gain and noise figure circles plotted on the source reflection coefficient plane of the Mitsubishi Electronics MGF-4300A transistor at 4 GHz.*

Example 9.3: The S-parameters and noise figure parameters of the Mitsubishi Electronics MGF-4300A transistor at 4 GHz, 2 volts and 7.5 mA are given in Example 9.2. Can a conditionally stable amplifier with the optimum noise figure be designed? Figure 9-2 shows the optimum noise match in the stable source match region of the Smith chart. If we match the input of the transistor with the optimum noise figure, the output reflection coefficient can be found using Equation 6.6.

$$\Gamma_0 = S_{22} + \frac{S_{21}S_{12}\Gamma_s}{1 - S_{11}\Gamma_s}$$

$$\Gamma_0 = 0.627@-58.3° + \frac{\left(3.434@106.2°\right)\left(0.072@36.5°\right)\left(0.740@62.0°\right)}{1 - \left(0.863@-79.1°\right)\left(0.740@62.0°\right)}$$

$$\Gamma_0 = 0.534@-100.0$$

The output matching circuit yields the conjugate of Γ_0 inside the load stability circle. The amplifier would not be unconditionally stable. The constant noise figure and constant gain circles illustrate two contours on the source reflection coefficient plane. This allows us to trade off noise and gain. Once the source match is chosen, the output reflection coefficient is calculated using Equation 6.6. The matching circuit should then be designed to match this reflection coefficient.

The design of broadband low noise amplifiers is similar to the design methods discussed in Chapter 7. The input-matching circuit should be a lossless, reactive type because any losses in the input will increase the noise figure of the amplifier. A model of the optimum noise match is just as useful as models in the input of the transistor used for broadband amplifiers. The optimum noise figure match of a Kukje KH1031-C02 FET from 4 to 12 GHz is plotted on a Smith chart in Figure 9-3. The frequency characteristic can be approximated by a series resistor, capacitor and inductor, as shown in Figure 9-4.

This model is useful in designing an amplifier with the lowest noise figure over frequency, neglecting the gain or stability of the amplifier. In practice, we plot stability, constant noise figure and constant gain circles at several frequencies across the band. We choose a source match at each frequency that gives us a flat gain and noise figure across the band. Then we calculate the output match that corresponds to the various source impedances across the band and check the stability of the transistor.

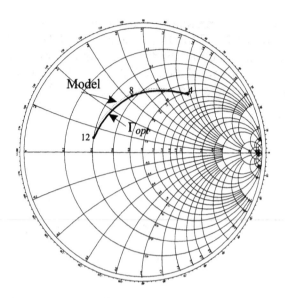

Figure 9-3 *Optimum noise figure of the Kukje KH1031-C02 FET plotted on the input-source reflection coefficient plane from 4 to 12 GHz.*

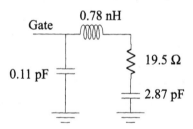

Figure 9-4 *Approximate model of the input of the Kukje KH1031-C02 FET at the optimum noise figure match.*

When the transistor is matched at the optimum noise figure at all frequencies of interest, the noise figure begins low at low frequencies and increase as the frequency increases. A constant noise figure and gain are usually desired for most broadband amplifiers. The highest optimum noise figure will occur at the highest frequency as well as the lowest associated gain. The input matching circuit should be designed to optimally match the transistor for noise figure at the highest frequency ($F_{min,h}$). At lower frequencies, the input should be matched with the constant noise figure circle corresponding to $F_{min,h}$. The output-matching circuit is designed to give the highest associated gain at the upper frequency band limit and then mismatch the device to achieve a constant, or flat, gain at lower frequencies. Some mismatch can be designed into the input- and/or the output-matching circuit to compensate for the 6 dB-per-octave gain roll-off.

9.3 Cascaded Noise Figure

The noise figure of cascaded networks depends on the noise figure and gain of the different stages. Figure 9.5 shows two circuits in cascade, each with a noise figure and gain. The ratio of signal to noise at the output of the first stage is

$$\frac{S_o}{N_o} = G_1\left(F_1 + \frac{S_i}{N_i}\right)$$

9.13

The signal-to-noise ratio at the output of the second stage is

$$\frac{S_o}{N_o} = G_2\left(F_2 + G_1\left(F_1 + \frac{S_i}{N_i}\right)\right)$$

9.14

When two circuits are cascaded, the total noise figure of the composite network is

$$F_t = F_1 + \frac{F_2 - 1}{G_1}$$

9.15

where F_1 and G_1 are the noise figure and gain of the first stage, and F_2 is the noise figure of the second stage. Amplifier combinations of three stages or more will have a noise figure of

$$F_t = F_1 + \frac{F_2 - 1}{G_1} + \frac{F_3 - 1}{G_1 G_2} + \frac{F_4 - 1}{G_1 G_2 G_3} + \cdots$$

9.16

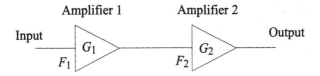

Figure 9-5 *Two amplifiers in cascade.*

The noise figure of the first stage has a large contribution on the overall system noise figure. The noise contribution of the second stage is reduced by the gain of the first amplifier. It is desirable to place the amplifier with the lowest noise figure first. If the gain of the low noise amplifier is high, the noise figure of the second stage amplifier can be neglected. However, since there is a trade-off between noise figure and gain, we have to be careful not to sacrifice too much gain to achieve low noise. If we do not pay attention to the correlation between noise figure and gain, the next component in the system could add a significant amount of noise and subsequently increase the overall noise figure of the system.

Example 9.4: Figure 9-6 shows a block diagram of a microwave receiver. A low noise amplifier with a gain of 15 dB and a 1.5 dB noise figure is at the input. It is followed by a mixer that has a loss of 6 dB and a noise figure of 8 dB. There is also a filter that has a loss and noise figure of 4 dB. An intermediate frequency amplifier with a gain of 32 dB and a noise figure of 4 dB is last in the cascade. What is the noise figure of the receiver?

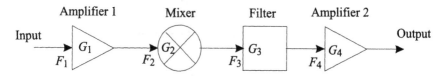

Figure 9-6 *Block diagram of microwave receiver.*

We convert the noise figure and gain from dB to noise factor and power gain.

	Amplifier 1	Mixer	Filter	Amplifier 2
Gain	31.62	0.25	0.40	1585
Noise Factor	1.41	6.31	2.51	2.51

We use Equation 9.16 to find the noise figure of the receiver.

$$F_t = 1.41 + \frac{6.31-1}{31.62} + \frac{2.51-1}{(31.62)(0.25)} + \frac{2.51-1}{(31.62)(0.25)(1585)} = 2.25$$

$$= 3.5 \text{ dB}$$

When a low noise amplifier is used in a system with other amplifier stages, it is desirable to have a large gain associated with the low noise amplifier. The second stage contribution to the system noise figure can be minimized when the gain of the first stage is high. There can be a significant increase in system noise if the gain of the first amplifier is low. Sometimes it is wise to design a low noise amplifier with a slightly higher noise figure to obtain a higher gain. Along with the noise figure and gain of the amplifier, *noise measure* is used to evaluate the performance of a low noise amplifier. Noise measure describes the noise figure of an infinite string of identical amplifiers in cascade. Continuing from Equation 9.16, an infinite cascade of amplifiers with noise figure of F and gain of G will have a noise figure of

$$NM = F + \frac{F-1}{G} + \frac{F-1}{G^2} + \frac{F-1}{G^3} + \cdots \qquad 9.17$$

which is an infinite series with the solution

$$NM = \frac{GF-1}{G-1} \qquad 9.18$$

The noise measure is calculated from the amplifier noise figure and gain.

Example 9.5: Calculate the noise measure of a Mitsubishi Electronics MGF-4300A matched for optimum noise figure. Is this the lowest noise measure for that transistor? Figure 9-2 shows that the gain at the optimum noise figure match is approximately 16 dB. The noise measure of an amplifier match for optimum noise figure is

$$NM = \frac{(39.8)(1.11)-1}{39.8-1} = 1.13$$

The same chart shows that the noise figure is approximately 1.18 at the maximum gain point. The noise measure at this point is

$$NM = \frac{(47.69)(1.18)-1}{47.69-1} = 1.18$$

In this case, the optimum noise match is very close to the optimum noise measure.

9.4   Low Noise Bias and Dynamic Range

The transistor bias voltage and current are usually different for the lowest noise figure and highest gain. Figure 9-7 shows a plot of the noise figure and gain as a function of drain current for the MGF-4300A at 10 GHz. The lowest noise figure occurs at a drain current of 8 mA with a gain of 10.5 dB. The maximum gain occurs at a drain current of 30 mA.

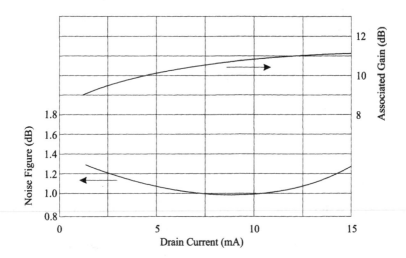

Figure 9-7 *The minimum noise figure and gain of the MGF-4300A transistor as a function of drain current at 12 GHz.*

The noise figure of an amplifier helps to define the lowest signal level that can be used. Another measured parameter helps define the highest signal level. A common parameter is the third-order intercept point, or *IP3*, which measures the nonlinearity of an amplifier. As input signal power increases, it will begin to overdrive the amplifier and eventually cause distortion. The output voltage of an amplifier can always be expressed as a power series.

$$V_o = a_o + a_1 V_i + a_2 V_i^2 + a_3 V_i^3 + \cdots \qquad 9.19$$

The coefficient a_o is the DC voltage at the output, and a_1 is the linear gain of the amplifier. When the input signal is small, the coefficients a_2, a_3, ... are very small, and the amplifier is considered to be linear. As the input signal becomes large, the higher order terms become larger, and the output becomes distorted. When the input signal is a single frequency sinusoidal voltage, the output will contain many upper harmonics.

This may not be a problem when the input is a single frequency. Most amplifiers need to operate in an environment with many signals or a signal that is spread over a certain bandwidth. Nonlinearity will mix the separate frequency components together, creating spurious output frequencies. Let us consider the case in which the input to an amplifier consists of two signals of equal power that are closely spaced in frequency.

$$V_i = A(\cos \omega_1 t + \cos \omega_2 t) \qquad 9.20$$

This is called a *two-tone* test. The output will consist of harmonics of the two input frequencies and their linear combinations, $m\omega_1 + n\omega_2$ or $m\omega_1 - n\omega_2$ where m and n are integers. The output contains signals of different orders where the order is equal to $|n| + |m|$. The first order signal is the one that is an exact replica of the input signal with gain (or loss). The second order is the second harmonics and signals at the sum and difference frequencies. The third order is signals at frequencies of $3\omega_1$, $3\omega_2$, $2\omega_1 - \omega_2$, and $\omega_1 - 2\omega_2$.

Third-order distortion products can pose a serious problem because their frequency is close to the original input signals. Figure 9-8 shows an illustration of the output spectrum of a nonlinear amplifier under a two-tone test. The third- and fifth-order intermodulation distortion products are also shown. The level of the third-order intermodulation distortion product will increase as the amplifier's input signal power grows. Figure 9-9 shows a plot of the output power versus input power of the first- and third-order products as they were measured on a spectrum analyzer. The third-order intercept point is shown on the graph. The first-order and third-order power plots are extrapolated with a straight line until they meet at the third-order intercept point.

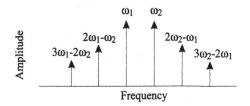

Figure 9-8 *The output spectrum of an amplifier with two signals on the input showing the third- and fifth-order intermodulation products.*

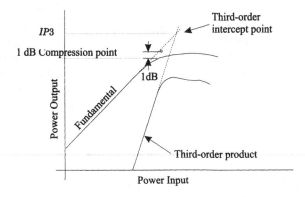

Figure 9-9 *A plot of input power versus output power in the first (fundamental) and the third-order intermodulation product as a function of input power.*

Also shown on the plot is the 1 dB gain compression point, which consists of the output power when the amplifier gain is one dB less than when the input power is small. The third-order intercept point and the one dB gain compression are both used to describe the upper limit of an amplifier's useful power range. Microwave systems have cascade rules for the third-order intercept point. The last amplifier or component in a system has the most influence on the third-order intercept point.

$$\frac{1}{IP3_{total}} = \frac{1}{IP3_n} + \frac{1}{IP3_{n-1}G_n} + \frac{1}{IP3_{n-2}G_nG_{n-1}} + \cdots + \frac{1}{IP3_1\left(G_nG_{n-1}\cdots G_2\right)} \qquad 9.21$$

The spurious free dynamic range (SFDR) denotes the difference between noise level and signal power when the third-order intermodulation product is equal to the noise in the detection bandwidth. The third-order intermodulation product increases 3 dB for every 1 dB increase in the two input signals. Theoretically, the SFDR would be zero at the third-order intercept point. For every 1 dB of reduction in output power, there is a 3 dB reduction in the third-order intermodulation distortion product. The result is a 2 dB improvement in SFDR. If P_N is the total noise power at the output of the receiver, the SFDR is

$$SFDR = \frac{2\left(IP3 - P_N\right)}{3} \qquad 9.22$$

where $IP3$ is the intercept point.

9.5   Summary

Many RF and microwave applications demand low noise amplifiers in the receiver. Amplifiers for these systems are designed to have a specified noise figure. This chapter has discussed a systematic method of designing amplifiers with a specified noise figure. Expanding on the information presented in Chapter 8, we discussed how the parameters that affect the noise figure of two-port networks apply to the design of amplifiers.

Section 9.2 explained how to design low noise amplifiers. Information on noise parameters, such as the minimum noise figure, optimum noise match, and the correlation resistance is provided by the transistor manufacturer. With these values and the S-parameters at the frequency of interest, the constant noise figure and available gain circles can be plotted on the source reflection coefficient plane. We find two families of circles, one of constant noise figure and the other of constant gain. We then can see which source reflection coefficients will give a certain combination of noise figure and gain. Sometimes this will not be necessary. We may want the lowest possible noise figure and be satisfied with the resultant gain at the optimum noise figure source match. When that source match falls into a region of instability or does not yield enough gain, it is necessary to find the constant noise figure and gain circles.

Section 9.3 showed how cascading two or more amplifiers affects the noise figure of the amplifier chain. The amplifier that is closest to the input has the most effect on noise figure. The next amplifier has a lesser effect, which depends upon the gain of the first stage and the noise figure of the second stage. The second stage has less effect on the noise figure of the system when the first stage has more gain. Often an amplifier that has been optimized for noise figure does not have much gain. In this case, the second stage may cause the noise figure of the combination to be too high. It may be necessary to redesign the first stage and give up some noise figure performance to obtain more gain. Noise measure is one indication of the balance between noise figure and gain. The noise measure of an amplifier is the noise figure of an infinite series of identical amplifiers connected in cascade. The difference between the noise figure and noise measure of an amplifier indicates how much the second stage will effect the system's noise figure. The larger the difference, the more effect will the second stage have.

Finally, Section 9.4 discussed the bias voltage and current needed to achieve the lowest noise figure, which is obtained at a low voltage and current through the transistor. This contrasts with higher levels of voltage and current needed to obtain the highest gain. Even higher levels are needed to achieve the highest power from the transistor. Careful bias control is essential to achieve the noise figure, gain and output power goals for the amplifier. The dynamic range of an amplifier is the difference between the noise level and the maximum power output from the amplifier. The output noise is determined by the noise figure and gain of the amplifier. The maximum output power is usually considered to be the one dB gain compression point.

9.6   Problems

9.1   Using the data in Table 9-1, plot the constant 0.85, 1.00, and 1.4 dB noise figure circles for the Kukje KH1031-C02 at 12 GHz.

9.2   Find the output reflection coefficient of the MGF-4300A in Example 9.2 when the input is matched to give a noise figure of 1.25 and a gain of 15.5 dB.

9.3   The S-parameters and noise parameters at 10 GHz for the NEC 76184A are shown below. Find the maximum possible gain that can be achieved by an amplifier with a noise figure of 2.5 dB. Find the input and output reflection coefficients for the amplifier.

$$S_{11} = 0.56@96° \quad S_{12} = 0.131@–29° \quad S_{21} = 1.58@–33° \quad S_{22} = 0.41@–166°$$

$$F_{min} = 2.15 \text{ dB} \quad R_n = 12.5 \quad \Gamma_{opt} = 0.41@–152.0°$$

9.4   Using the S-parameters and noise parameters of an Avantec 10600 GaAs FET at 6 GHz, find the input and output matching circuit reflection coefficients that give an unconditionally stable amplifier with the lowest possible noise figure and the most gain at that noise figure.

$$S_{11} = 0.86@–59° \quad S_{12} = 0.101@64° \quad S_{21} = 2.27@120° \quad S_{22} = 0.57@–17°$$

$$F_{min} = 1.1 \text{ dB} \quad R_n = 54 \quad \Gamma_{opt} = 0.67@52°$$

9.5.   Calculate the noise figure of the microwave receiver in Example 9.4 again using a gain of 10 dB for the first low noise amplifier. Repeat the calculation setting the gain of the low noise amplifier to 20 dB.

9.6.   Repeat Example 9.5, assuming the transistor has a MSG of 19 dB instead of 16.8. Use the same noise figures and gain at Γ_{opt} as in the previous example.

9.7.   The components in the receiver shown in Figure 9-6 have third-order intercept points of $IP3_1 = 20$ dB, $IP3_2 = 10$ dB, $IP3_3 = 1000$ dB, and the last stage has an $IP3_4 = 30$ dB. Using the component gain and loss given in Example 9.5, find the third-order intercept point of the entire receiver.

9.7   Appendix

Cnoise.m

```
function[c,r]=cnoise(f,fmin,gopt,rs)
% This function calculates the constant noise figure
% circles.
% Usage: [c(n),r(n)]=cnoise(f(n),fmin,gopt,rs)
%    where f(n) is a vector holding a list of noise figures
%         n is the number of circles wanted
%         fmin is the minimum noise figure
%         gopt is the optimum noise match
%         rs is the noise resistance/impedance
```

```
%        c is the center of the circles
%        r is the radius of the circles
j=sqrt(-1);
temp=size(f);
n=temp(2);
for ii=1:n
  if (f(ii)>fmin)
    bign=(f(ii)-fmin)*abs(1+gopt) ^ 2/(4*rs)
    c(ii)=gopt/(bign+1);
    r(ii)=sqrt(bign*(bign+(1+abs(gopt) ^ 2)))/(bign+1);
  end
end
```

10

Summary

10.1 Amplifier Design Review

In the previous chapters, we provided tools and techniques for designing stable amplifiers for high-frequency applications. These amplifiers are intended for applications in which the signal is *small* and the amplifier circuit is *linear*.

Chapter 2 introduced networks. Simple current and voltage sources can be used to analyze circuits and describe their signal response. This technique assumes that the amplitude or strength of the signal does not affect the circuit's operation. Z-, Y-, and chain parameters can be used to characterize a circuit. Many small networks can be connected in series, parallel, and cascade to build large networks. These parameters are commonly used in computer-aided design (CAD) programs.

Chapter 3 introduced the properties of transmission lines and high-frequency circuits. Transmission lines play a crucial role in guiding signals from one place to another and within the amplifier circuit itself. We have presented the fundamentals of signal propagation in a transmission line. Voltage and current vary in time and position in a transmission line as signals propagate in them. It is often awkward or might even be impossible to measure these voltages or currents directly. Instead, these signals are characterized by power waves, which propagate in both directions along the transmission line. S-parameters describe the network response to these power waves. They are valuable because they relate directly to power flow, which is easier to measure at high frequencies than voltage and current. S-parameters for high-frequency transistors are provided by the manufacturer. High-frequency transistors are the key elements in microwave amplifiers.

In Chapter 4, we showed how to use S-parameters to analyze networks. Methods of analyzing interconnected networks are needed to solve many measurement and design problems. The two primary tools are mapping functions and signal flow diagrams. Mapping functions are useful when analyzing the effect of terminating one port of a network. Terminating one network terminal affects the input impedance at other ports. Mapping functions provide a set of equations that describe how this effect unfolds under certain conditions. For example, if one port is terminated with a mismatch at an unknown phase, we can determine the range of reflection coefficients that could occur at the other ports. We presented signal flow diagrams for the analysis of one or more multi-port networks connected together. Signal flow diagrams are an effective tool for analyzing interconnected networks.

Chapter 5 introduced impedance-matching circuits, which, by matching impedances, ensure that the power transfer between impedances is efficient. Commonly, power is delivered from the source to the transistor's input and then extracted from the transistor's output into a load. We focused on eliminating the reflection coefficient between a source and load. We have presented analytical methods of designing narrowband matching circuits using lumped elements such as capacitors and inductors, and transmission lines connected in series or as open or shorted stubs. Several different circuit topologies and sample programs were supplied to expedite the design process.

Chapter 6 discussed the design of narrowband amplifiers. Stability is an important consideration in amplifier design. If an amplifier circuit oscillates, it will not have the desired gain and impedance match. Each new amplifier design begins with the calculation of the stability factor K. If the stability factor is less than one, the input and output stability regions must be found. We showed how to determine the input and output match that gives the highest possible gain, i.e., the conjugate match points. With these values of reflection coefficient, the synthesis methods of Chapter 5 can be used to design the input and output matching circuit. If the conjugate match falls within a region of instability, another match point has to be chosen. After describing the different forms of power gain, we derived constant gain contours in the reflection coefficient plane from the power gain equations, which allowed us to choose a gain that provides a comfortable level of stability.

Chapter 7 gave a brief introduction to broadband amplifier design. Various transistor models and some examples of models that match real data were discussed. These models are useful because many broadband design techniques utilize elements of the transistor model in the matching circuit. For example, if the transistor model has a series inductor as the first component in the model, it can be absorbed into the matching circuit itself. We then showed the theoretical limit on bandwidth and gain. Finally, a few design methods were mentioned, including impedance transformations, balanced amplifiers, distributed amplifiers, active matching, and lossy matching circuits.

In Chapter 8 we considered noise and noise figure, the source of noise and how it can be modeled. Noise, as was shown, is modeled as equivalent voltage and/or current sources with internal impedances. The noise figure of a circuit the degree to which the signal-to-noise ratio degrades as a signal passes through it. Further, the amount of noise power delivered depends on the load impedance, and therefore the noise figure also depends on load impedance. We concluded Chapter 8 with a short discussion of noise figure measurement.

Chapter 9 demonstrated how gain and noise figure are used to design amplifiers. In most cases, low noise and maximum gain cannot be achieved simultaneously. We showed how to calculate contours of constant noise figure along with the constant gain circles, demonstrated how a trade-off between noise figure and gain can be achieved and explained the noise figure of receiver systems. Each stage of a receiver or amplifier chain contributes to the overall noise figure. It is important to know how much contribution is due to each stage in the chain. At the end of Chapter 9, distortion products were introduced as well as the concept of third-order intercept point (IP3). The IP3 of an amplifier provides information on the approximate range of linear operation and the maximum spurious-free dynamic range.

10.2 Nonlinear Devices

Microwave systems also use nonlinear devices, which convert one frequency to another and boost a signal's power. Some of the most common nonlinear circuits are power amplifiers. The amplifiers we designed in this book are for small signals and operate well within Class A. The efficiency, measured with the ratio of RF power gained by the supplied DC power, is very small. The efficiency of power amplifiers is maximized to keep the power dissipated in the transistor at a minimum, which usually means operating in the nonlinear Class AB. As a result, the S-parameters change and now depend on the matching circuits as well as the input signal power.

Another common component is the frequency mixer. The ideal mixer multiplies two signals. When two cosine waves are multiplied, the results are two cosine waves consisting of the sum and difference of the two frequencies. This is a nonlinear operation. Regenerative receivers have one mixer where the received frequency is mixed with a local oscillator (LO), and the difference frequency is selected for demodulation. A superheterodyne receiver uses two such mixers to downconvert the signal to a lower frequency. The output of a real mixer is a collection of frequencies that are the sum and difference of the LO and receive signal harmonics. The largest frequencies produced by a mixer are the input frequency, the LO frequency, their sum, and their difference. To build a 10 GHz receiver, we can use the 10 GHz signal and mix it with a 9 GHz LO. A low pass filter will pass the 1 GHz difference frequency and block all the others. The 1 GHz intermediate frequency (IF) has the same information as the original 10 GHz signal. Processing the 1 GHz signal becomes easier with off-the-shelf components.

The amplifiers designed in this book have avoided the potentially unstable regions of the reflection coefficient plane. Oscillators are designed to be unstable. The transistor becomes unstable by matching circuits, which usually have a resonant element to select the frequency of oscillations. Oscillators require special design techniques. However, many amplifiers can be unintentional oscillators.

Power detectors are commonly used in systems as receivers and test equipment. A detector emits a voltage that is proportional to the input signal power. Amplitude modulated signals are demodulated using a detector. Signal power may be monitored so that an automatic gain control (AGC) loop can be implemented. In test equipment, reflection coefficient and gain are measured with directional couplers and power detectors. Finally, detectors are used to monitor local oscillator power or signal power for an operator or technician.

10.3   Passive Components

High-frequency systems also use passive components such as attenuators, which accurately set the power level of a signal or LO. Attenuators are also used to isolate components. For example, when two amplifiers connected in cascade are not perfectly matched, the reflection coefficients could be phased in such a way that reinforces the standing waves between them. This could cause dips in the frequency response of the amplifier chain. The addition of a 2 dB attenuator between stages increases the isolation between them by 4 dB.

Another component used to prevent interaction between amplifiers is the isolator. Isolators allow signal flow only in one direction. An isolator is a special form of a circulator, which has three ports. In a circulator, signals flow from the first port to the second port, from the second to the third port and from the third to the first port. They do not flow against this circular pattern. Terminating the third port of a circulator makes an isolator. Power entering the first port flows to the second port. Power flowing into the second port flows into the load on the third port. If the load is perfectly matched, there will be no power incident on the third port and no power will leave the first port.

Filters are used to select the receive band and the difference frequency output from a mixer. Chapter 6 described filter synthesis as a broadband matching technique. Filters are constructed in many different ways and include cavity filters, lumped element filters, standing acoustic wave (SAW) filters and many other types. Filters are commonly combined to build components such as duplexers. Duplexers split signals into two bands. Signals in one band exit through one port, and signals in another band exit from another.

10.4 Other Microwave Specialties

Microwave amplifier design requires special methods and parameters. All techniques shown in this book have been in the frequency domain. The analysis and synthesis techniques are single-frequency methods. Sometimes special nonlinear models are used with a time-domain CAD program, where the response of the circuit is numerically evaluated at discrete time steps instead of frequency steps. A large signal model may have capacitors and resistors that change values in response to changes in voltage or matching impedance. Power amplifiers and oscillators are examples of large signal applications. When instability is created in a transistor, oscillations increase until the nonlinear behavior at large signal levels prevents any further increase.

Filters are used in nearly all RF and microwave systems. The most common filter types are bandpass and lowpass filters. On occasion, there may be a requirement for a highpass or bandstop filter. Filters can have many electrical characteristics depending on the passband, stopband, and phase delay characteristics that are desired. The filter's construction is highly dependent on the frequency of operation, the input transmission line type, and the Q of the resonant elements in the filter. Some filters are made by connecting resonant cavities together. Others are lengths of microstrip line that are physically connected or electromagnetically coupled. Some have special ceramics that are precisely shaped to make them resonant at the correct frequencies.

Advancements in integrated circuit technology continue to push the upper frequency limit on monolithic circuits. Monolithic microwave integrated circuits (MMIC) are fabricated on a semiconductor chip. Capacitors are consist of a thin dielectric between two metal layers. Inductors are made by printing spiral traces of metal on the surface of the semiconductor. Doping portions of the semiconductor produce resistors and active devices, such as transistors and diodes. The circuits have to be small to maximize the number that can be placed on a wafer of semiconductor material.

Lumped elements are preferred because they are usually smaller than transmission lines. Monolithic components are based on models that describe their behavior over a broad range of frequencies. The cost of making prototype circuits is high and the inability to tune the circuit after it is fabricated are two drawbacks that should be considered before a design is realized.

Microwave and millimeter-wave electronics rely on special materials and components. The design and process control of these materials includes circuit board material, ceramics, metal alloys, semiconductors, bonding materials, radiation absorbers, coaxial connectors, and many others. The best way to become familiar with these items is to work with an experienced design engineer and to consult articles and advertisements in technical journals and magazines.

Index

Printed in the USA
CPSIA information can be obtained
at www.ICGtesting.com
JSHW011451221024
72173JS00005B/1024

9 781884 932069